I0503253

The Limitless Sky

Air Force Science and Technology Contributions to the Nation

Edited by:

Alexander H. Levis
Chief Scientist, USAF

with the assistance of
Colonel John C. Bedford, USAF, Ret.
Captain Sandra Davis

Air Force History and Museums Program
United States Air Force
Washington, D.C.
2004

"...problems never have a final or universal answer, and only constant inquisitive attitude towards science and a ceaseless and swift adaptation to new development can maintain the security of this nation through... air supremacy"

Dr. Theodore von Kármán to General Henry H. "Hap" Arnold
15 December 1945

To the Air Force scientists and engineers, military and civilian, for keeping true to von Kármán's vision.

Acknowledgement

Neither the Symposium nor this book would have been possible without the support of a number of organizations. The Air Force Association provided a venue for the Symposium and made it a part of its Fall '03 convention in Washington, D.C. The Air Force Legislative Liaison and the Public Affairs Offices provided administrative support and publicity, while the Centennial of Flight Office included this activity in the centennial program. The organizing committee was chaired by Col. John B. Bedford, AF/ST, and included Alexander G. Levis and Nancy C. Stenger from AF/ST, Bill Heimdahl from AF/HO, Col. Tom Schubert from AF/LL, Maj. Michelle Ryan, AF/CVAH, and Napoleon Byars from AFA. The Program Committee had the responsibility of selecting the topics and working with the multiple authors to produce the papers and the presentations. The committee was chaired by Alexander H. Levis, AF/ST, and included Barbara A. Wilson, AFRL/CT, Col. Donald R. Erbschloe, AFOSR/CD, George W. Bradley III, AFSC/HO, Ruth P. Liebowitz, ESC/HO, and Maj. John Beaulieu, AF/HO. Thanks are also due to several individuals who introduced the papers at the Symposium: LtGen Tom Hobbins, AF/XI, Maj Gen Don Lamberson (Ret.), Barbara Wittig, Jacob Neufeld, and Richard Wolfe, of AF/HO for editorial assistance and Susan Linders, of Air Force Media Services for design and layout.

Content

The Limitless Sky

The old dream of mankind to fly and reach the sky—as exemplified by Daedalus and Icarus of ancient Greek lore—became a reality on December 17, 1903, at Kitty Hawk, North Carolina, and ushered in a period of unprecedented technological development. From Orville Wright's short flight a few feet off the ground, to the commercial airliners flying at 37,000 feet, the limits of the sky have receded farther and farther away. In the second half of the twentieth century, powered flight extended to satellites orbiting the earth, to landings on the moon, and probes that explore our planetary system. The Hubble space telescope has been producing breathtaking pictures of distant galaxies not observable from terrestrial telescopes. The sky has become limitless.

The Air Force, from its very beginnings as the Army Air Corps, has contributed significantly to all aspects of powered flight that have not only enhanced the defense capabilities of the United States, but also produced a broad spectrum of non-military applications that have improved the quality of life throughout the world. Consequently, to celebrate the centennial of flight, a one-day symposium was held on September 17, 2003, in which the stories of some of the contributions, and of the people who made them, were told. The objective of the symposium was to present technological developments that have produced new capabilities or opened new ways for achieving objectives. Each paper is the result of a collaborative effort of historians, who have placed the contribution in its historical perspective; technologists, who have described the essence of the scientific or technological contribution; and Air Force senior officers, who have shared their personal experiences on how that technological development affected operations or missions.

The nine papers included in this volume were selected because of their diversity and because they illustrate clearly several key themes. First, it takes a long time from the onset of a new idea to the production of a useful product that enhances operations, something on the order of twenty years. One has to believe in the idea and stay the course, in the face of adversity, to obtain results. Consistent, steady funding is a must. Second, research results rarely lead to what was envisioned in the beginning as a relevant application. Indeed, research in atomic clocks enabled the Global Positioning System (GPS), but that was not the motivation for the research. Similarly, early research on lasers hardly anticipated the proliferation of commercial products or at-home entertainment via DVDs. Third, it usually takes a confluence of several disparate developments to produce a new capability. A vibrant, interacting scientific and engineering community is essential to achieve the breakthroughs that will continue to provide the nation with air supremacy.

The first paper tells the story of GPS. It is not only a story of scientific and technical achievement, but also one of suspense that illustrates the difficulties of introducing revolutionary technologies. The failures and successes in that program attest to the vision and perseverance of the people who made it happen. The GPS finally solved the problem that had challenged seafarers since antiquity. It has also enabled a multitude of civilian and military applications.

The second paper looks at a development that has revolutionized the employment of air power. The use of laser research and the GPS has enabled ever-increasing precision strikes at ever-increasing distances. From tens of bombers dropping hundreds of bombs to hit a single target in World War II, now a single aircraft can strike several targets. Precision also raises expectations for minimizing collateral damage and puts high priority on detecting targets, identifying them correctly, and tracking them. Advances in electro-optical/infrared imaging, in radar, and in lasers have become the triad for remote sensing, where the objective has been to "overcome the problems of great distances, weather, and darkness," articulated by von Kármán. The third paper focuses on infrared cameras from their first beginnings to their employment in unmanned aerial vehicles such as the Predator and Global Hawk. The fourth paper looks at airborne radar, whether in AWACS or in Joint STARS, two systems that have revolutionized the conduct of air-to-air and air-to-ground operations.

Soon after the first powered flight in 1903, the effects of the high altitude environment on the performance of pilots became a concern. Research on the effects of hypoxia went back to balloonists in the late nineteenth century. With the creation in 1918 of the Air Service Medical Research Laboratory, it became possible to establish the scientific foundation for the development of a wide variety of equipment, from oxygen masks to the pressure suits of today, that would enable air crews to fly higher and for longer periods of time. The fifth paper traces that history and the accomplishments of such pioneers as Armstrong, Stapp, and Kittinger. The paper also includes the story of a particular mission of the SR–71, an aircraft that flew at 80,000 feet, at the edge of space, at three times the speed of sound—an extreme environment indeed.

The sixth paper, not presented at the symposium, describes the Air Force research in weather, not only the effort to understand terrestrial weather, but also space weather. The launching of U.S. satellites, which started in 1958, led to the discovery of the Van Allen belts. This was a major scientific discovery that began the mapping of the space environment, a necessary condition for sending unmanned and manned spacecraft to space.

To reach space, whether for scientific, military, or commercial purposes, a launch capability was needed. In the early years of the space age, this capability was provided by the Air Force's long-range ballistic missiles—Atlas, Titan, and Thor. The seventh paper records the history and the systems that have allowed the U.S. to launch satellites that have provided extraordinary sensing capabilities and global communications. The evolution of military satellite communications from concept to reality, an indispensable capability for today's Air Force, is documented in the eighth paper.

Finally, as we start the second century of powered flight, the Air Force is once more expanding its capabilities using advanced technology. The wave of the future is directed energy that is expected to revolutionize the science and art of warfare. Since the invention of the laser in the 1960s, the Air Force has been conducting research and development for an operational platform that would use lasers to destroy ground or air launched missiles. The Airborne Laser (ABL), now in development, is the perfect example of the need for a number of technologies to work together – lasers, adaptive optics, and control engineering – to make a new capability a reality.

These nine papers present only a few of the milestones in this limitless journey of discoveries that started at Kitty Hawk and in which millennia-old limits were put aside. It was a journey in which the Air Force science and technology enterprise has made immeasurable contributions to the development of air power, to the nation's science and technology base, and to mankind's understanding of our world.

Alexander H. Levis, Sc.D.
Chief Scientist, USAF

15 March 2004

Precision Timing, Location, Navigation:
GPS and the Precision Revolution

Michael I. Yarymovych

Bradford W. Parkinson

George W. Bradley III

Brigadier General Daniel J. Darnell, USAF

Ivan A. Getting

Abstract

The history of human navigation extends thousands of years, from the invention of the compass needle, sextant, and chronometer, and arriving at the introduction of radio navigation in the 1920s and its operational use during World War II. Nuclear submarine operations created the need for satellite-based navigation, which then was further driven by the requirements of fast-moving airplanes. In the 1970s, the Global Positioning System (GPS) emerged as a radical new way to provide precise navigation for all of the U.S. armed forces across the globe. The development of GPS, the most revolutionary navigation tool since the invention of the chronometer, is described, and the technologies that went into its successful implementation are highlighted. Declared operational in 1995, it had been exploited while a developmental system by both civilian and military users for more than ten years.

Military use of GPS got its first operational application in Desert Storm, was then used for precision warfare in the Kosovo operation, and eventually became the major force-enhancer in Afghanistan and Iraq. The number of operational receivers has increased exponentially over the last decade as the technology has moved in diverse and unexpected directions. Although GPS was originally designed for defense missions, civilian receivers now far outnumber military receivers. GPS has become a global utility with immeasurable potential benefits for all humanity, thanks to military science and technology development and Air Force implementation.

HISTORY OF NAVIGATION

Advancing human civilization is motivated by three basic drives: to trade, fight enemies, and explore the unknown. To accomplish these basic objectives, humans have always sought ever better means to know where they were and where they were going. For 6,000 years, they have been developing ways to navigate to remote destinations. Driven mostly by the desire to transport goods by ship, early navigators, such as the Phoenicians, remained within sight of land, using a technique known as piloting that relied on navigators' recognition of coastal features. The magnetic compass was one of the most important inventions in the history of navigation. It first appeared in China around 1100 AD, and in Europe, approximately a century later. When forced to sail beyond landfall or in inclement weather, mariners tracked their position by dead reckoning, determining their location on the basis of speed, time, and direction, using the magnetic compass. This device has remained an important navigational tool over the years. Early aviators navigated from one town to another by using landmarks. Antiquity also offered another navigational system independent of terrestrial objects: celestial navigation. This system used the observed motions of the sun and stars. Its effectiveness increased over the centuries with advances in instrumentation, such as the astrolabe, sextant, and accurate portable timepieces.

Marine pilots would record their heading and distance traveled by hourglass, timing the passage of wooden logs thrown off the bow. Needless to say, this technique was notoriously inaccurate. The development of a sextant by 1731 (early versions existed in the thirteenth century) made determining latitude fairly routine. But the longitude problem was so vexing that England established a Board of Longitude in 1714 and offered a King's ransom of 20,000 pounds sterling to whoever could resolve it. Some of the greatest minds of Europe joined the race for a solution. Some believed that variations in Earth's magnetic field held the key; others insisted on celestial techniques. John Harrison, not a scientist but an artisan clockmaker, trumped them all by building a chronometer that, during long sea voyages, lost less than one second per day. The board was reluctant to confer the award on an individual who was not a member of the established scientific academy, and Harrison, greatly embittered, had to wait until 1763 to collect his prize. Harrison's chronometer gradually gained favor, but ironically, the pace of navigational advancement slowed during the industrial era. Still, the ability to move and communicate over long distances by telegraph and railroad spurred the need for civil and military time coordination. Thus, in 1884, at the height of the British Empire, Greenwich, England, was established as the world's Prime Meridian. Previously, each major nation established its own prime meridian and local time; the promulgation of Greenwich Mean Time abolished this practice and standardized navigational readings throughout the globe.

At the turn of the twentieth century, Guglielmo Marconi successfully transmitted radio waves across the Atlantic. The earliest radio navigation systems were based on the ability of a radio receiver with a loop antenna to determine the direction of a radio signal and its relative bearing to the transmitter. Over the years, use of radio signals has become much more sophisticated. During World War II, the British introduced the Gee system, and scientists at the Massachusetts Institute of Technology Radiation Laboratory developed the Long Range Navigation system, or Loran, which measured the difference in time of arrival of signals from synchronized pairs of transmitters at different locations. To provide better world coverage with fewer transmitters, a low-frequency system called Omega was developed. Omega used continuous wave radiation rather than pulses, as the original Loran did, and it gauged the difference in time intervals from ground stations by measuring the relative phase angles of transmitters from pairs of stations. Unfortunately, the accuracy of Omega was limited. To increase accuracy, Loran-C, using ground-wave propagation, was developed for tactical aircraft. It could accurately determine position to about 100 to 200 meters, compared to Omega which had an accuracy only within 2,200 meters. Loran-C was widely used during Vietnam. Unfortunately, all the early Loran systems, as well as Omega, were essentially two-dimensional. In effect, they located by latitude and longitude but not altitude, the third dimension. Today, a modernized version of Loran has become a worldwide standard for aircraft and coastal marine navigation. In the 1950s and 1960s, scientists developed inertial navigational systems for aircraft and missile systems. Inertial navigation uses neither landmarks nor celestial observations to determine location. It relies on internal systems such as gyros for calculating speed, distance, and direction to determine location. Nonetheless, what all these systems have in common is that they provide the

user with location information, and military theorists from Sun Tzu to Clausewitz have emphasized the importance of knowing the exact location of friendly and enemy forces.

While ground-based radio waves proved an important innovation in navigation, some problems were noted. Low-frequency radio waves are not easy to modulate and are subject to errors due to ionospheric factors and weather turbulence. High-frequency radio waves are limited to line of sight. This necessitates many fixed-site transmitters. Moreover, it was impractical to place a fixed-site transmitter at sea. Like ground-based radio navigation, celestial navigation was also problematic. Mariners had traditionally been faced with a limited ability to use celestial navigation during periods of intense cloud cover or dense fog. What was needed was a way to receive radio waves from a fixed point in the sky.

Before the launch of Sputnik in 1957, scientists were attempting to develop a system to track future U.S. satellites. In the mid-1950s, the Naval Research Laboratory (NRL) had proposed a system called Minitrack, which was built to track the movement of the Navy's planned Vanguard satellite and other early man-made orbiting objects by using the signals that they transmitted. Passive satellites, those that emitted no signal, however, were beyond Minitrack's capabilities, and a different methodology was needed to deal with them. Roger I. Easton, Don Lynch, Al Bartholomew, and others at the NRL began working on a system to track such passive objects. They moved an FM transmitter to Fort Monmouth, New Jersey, and used it to illuminate the Sputnik satellite when it passed over. This experience led them to conclude that if they could radiate a fan-shaped, continuous-wave beam, they could detect anything that passed through it. This was the origin of the Naval Space Surveillance System. Easton's group then turned its attention to devising a navigation system based on information obtained from orbiting satellites. At NRL they realized that a satellite navigation system could be established using satellites having synchronized clocks transmitting signals to users on the ground, who could synchronize their own clocks to those in the satellites. The range measured to each satellite produced a line position, just as if one had obtained a sextant sighting on a star. Essentially, Easton's system was still based on celestial navigation. He substituted a satellite in a predicted and measured orbit for a star, and synchronized clocks for chronometers that seafarers had used since the eighteenth century.

The successful launch of Sputnik allowed a practical test because it emitted an active signal. Researchers at the Johns Hopkins Applied Physics Laboratory (APL) had carefully studied Sputnik's radio signal and noted certain regular features. The most interesting of these features was the Doppler shift as the satellite passed overhead. This effect was caused by changes in the length of the line of sight and was enhanced by the satellite's high speed and low altitude. These scientists developed a computer program to determine Sputnik's orbit. Dr. Frank T. McClure of APL realized that the problem could be turned on its head: the process could be reversed. By measuring the Doppler shift to a satellite of known orbit, listeners could calculate their own positions. This solved an important problem for the

U.S. Navy, since it needed precise, all-weather positions for submarines and other ships. After speedy approval, a satellite navigation program, the Navy Navigation Satellite System (also known as Transit from its initial birth as Project Transit), was initiated under APL's management. The first two developmental Transit satellites were launched by 1960, and the system became operational by 1964.

Transit eventually deployed an operational constellation that included five polar orbiting satellites. They produced fixes every 35 to 100 minutes and provided horizontal accuracies of 100 meters or better for a stationary user. A moving receiver could compensate for velocity with some degradation in accuracy. Transit was not generally used by aircraft because of its incompatibility with the rapid platform motion of an aircraft. Additionally, aircraft require the third dimension, altitude, which the Transit system did not provide. Transit was, however, an important predecessor to the Global Positioning System (GPS) and pioneered a number of key technologies and concepts. Transit led to a great refinement of Earth's gravity-field model, successfully tested dual-frequency correction techniques for ionospheric-induced delays, and was crucial in developing stable and reliable frequency sources. Transit provided only periodic updates, and degradation for a moving user made it unsuitable for aircraft. By the late 1960s, better systems were being explored by the Navy. The system was renamed the Navy Ionospheric Monitoring System when it ceased to provide navigational information on 1 January 1997.

Another key program developed by the NRL was Timation. The goal of Timation was to orbit very accurate clocks. These clocks were to be used to transfer precise time among various laboratories around Earth. Under certain circumstances, users could also determine their positions by using the Timation signal. The approach was somewhat different from Transit in that the radio signal allowed direct ranging by using a technique known as side-tone ranging. The research that NRL conducted for Timation played a key role in developing the atomic clocks used on GPS.

While the Navy pursued Transit and Timation, the Air Force was also seeking better ways to determine location. Transit was less useful for the Air Force because the air service needed precise location on systems like rapidly moving aircraft and missiles. In 1960, the Raytheon Corporation proposed to the Air Force a concept for a three-dimensional type of Loran called the Mobile System for Accurate Intercontinental Ballistic Missile Control, or MOSAIC. Basically, MOSAIC used four continuous-wave transmitters at somewhat different frequencies, with their modulation locked to atomic clocks and synchronized via communication links. During flight, a missile would continuously compute its position by using signals from MOSAIC. Through this application, guidance of the missile would be less dependent on precise knowledge of the launch point. A central figure in this proposal was Dr. Ivan Getting, the first president of The Aerospace Corporation. Before assuming that position, however, Dr. Getting had been vice president of engineering and research at the Raytheon Corporation at the time they made their proposal. Many of the personnel involved in its conceptual beginnings, including

Getting, were now working for The Aerospace Corporation. This was an important factor, because the Department of Defense (DoD) had selected the Air Force as its executive agent for space launch, and the Air Force had contracted The Aerospace Corporation to be its system engineer. Over the years, these two figures, Ivan Getting from the Air Force and Roger Easton from the Navy, would play crucial roles in the evolution of the U.S. satellite navigation system

HISTORY OF GPS

One of the precursors to the GPS was an Aerospace Corporation and U.S. Air Force effort designated Program 621B and managed by an office in the Advanced Plans group at the Air Force's Space and Missile Systems Organization (SAMSO) in El Segundo, California. The Aerospace Corporation's Dr. Getting strongly advocated 621B. This program evolved directly into GPS, though not before significant modifications were made to the original Air Force only concept. By 1972, 621B had already demonstrated operation of a new type of satellite-ranging signal based on pseudorandom noise (PRN). Successful aircraft tests had demonstrated the PRN technique using ground-based "simulated" satellites located on the floor of the New Mexican desert. The PRN modulation used for ranging was essentially a repeated digital sequence of fairly random bits that possessed certain useful properties. The sequence could be generated by using a shift register or, for shorter sequences, could be stored in very little memory. Given the limited capabilities of contemporary computers, this was a crucial feature. A navigation user could detect the phase, or start, of the signal sequence and use this for determining the range to the satellite. The PRN signal also has powerful noise-rejection features and can be detected even when its power density is less than one-hundredth that of ambient radio noise. Furthermore, all satellites could broadcast on the same nominal frequency because properly selected PRN codes were nearly orthogonal.

When locked on to a particular PRN sequence, all other PRN sequences appear to the user as simple noise. The PRN sequence can be tracked even in the presence of large amounts of noise, so other signals on the same frequency do not generally jam the signal of interest. The ability to reject noise also implied a powerful ability to reject most forms of jamming or unintentional interference. In addition, a communication channel could be included by inverting groups of the repeated sequences at a slow rate (50 bps is used in GPS). This communications channel allowed the user to receive the ephemeris, clock, and health information directly as part of the single navigation signal. The original Air Force concept visualized several constellations of satellites in highly elliptical orbits with 24-hour periods. This constellation design allowed deploying the satellites gradually (for example, to cover North America first), but it complicated signal tracking because of the very high line-of-sight accelerations. Initially, the concept relied on continuous signal generation on the ground with continuous monitoring and compensation for ionospheric delays.

In 1969, the Air Force awarded contracts to four companies—TRW Systems, Magnavox Research Laboratories, the Grumman Aerospace Corporation, and the Boeing Company—to refine the design and determine a cost for the proposed 621B navigation system. During 1971 and 1972, tests of operator equipment using ground and balloon-carried transmitters at White Sands Proving Ground achieved accuracies to within 50 feet. However, the DoD, because of service concerns, would still not commit to the Air Force program. A new figure emerged who would provide a solution to the deadlock. In late 1972, General Kenneth Schultz, then commander of SAMSO, appointed Colonel (Dr.) Brad Parkinson, one of the authors of this article, as the Air Force 621B program manager and directed him to gain approval for the concept-validation phase of the Defense Navigation Satellite System, as the new DoD satellite navigation system was originally known. After many briefings to senior personnel in the Pentagon, a Defense Systems Acquisition Review Council meeting was held in August 1973, at which Dr. Parkinson presented a brief on the Air Force 621B program. The review council initially denied Parkinson approval to proceed with the program.

Subsequent to the acquisition review council briefing, Dr. Parkinson presented the concept to the Air Force Chief Scientist, Dr. Michael Yarymovych, and the Director of Defense Research and Engineering, Dr. Malcom Currie, both of whom quickly appreciated the value of a three-dimensional, continuous, 10-meter positioning system. However, they thought the reason the program failed to win approval from the acquisition review council was that it was not a truly joint endeavor, incorporating the best technology and concepts across DoD. As a consequence, Dr. Currie essentially asked Dr. Parkinson to go back to the drawing board. Acting on that guidance, Dr. Parkinson assembled about ten of his key program members in the halls of the Pentagon during the hot Labor Day weekend of 1973. The end product of their labors was an all-encompassing synthesis system concept that would later be named the Global Positioning System. By mid-December 1973, senior DoD officials had been briefed, and a reconvened Defense Systems Acquisition Review Council granted approval. By June of 1974, the satellites, ground control system, and user equipment were on contract.

The Air Force Chief of Staff General John D. Ryan directed Air Force Systems Command to establish a joint program office at SAMSO to manage the program. All three services, as well as the Marine Corps and the Defense Mapping Agency, participated. The Air Force and Navy had reached a compromise program that used elements from both the Navy and Air Force systems: the Air Force's signal structure and frequencies, and the Navy's satellite orbits. The system would also use atomic clocks, which the Navy had already successfully tested in its Timation program. The initial four-year validation portion comprised a four-satellite configuration. In a surprise bid, Rockwell International won the first development contract. The first GPS satellite launch in February 1978 led to successful validation of the concept (Figures 1, 2).

Because the Air Force was not comfortable with having to shoulder the entire financial burden for the program, it attempted to cancel GPS several times. Although Air Force Chief of Staff General Lew Allen was a strong advocate of GPS, the civilian leadership was responsible for keeping the program alive. It took twenty

Figure 1. A Phase I GPS satellite built by Rockwell (now Boeing)

Figure 2. The primary booster for GPS satellites is the Delta II rocket

years, but finally, in March 1994, the launch of the twenty-fourth Block II satellite completed the GPS constellation. GPS was declared operational in December 1995, although civilian and military users had been using the available developmental system for more than ten years. The subsequent operational satellites incorporated certain additional non-navigation payloads, which enhanced their value but also delayed full operation. Without a doubt, the combination of near-concurrent efforts of the Navy and Air Force, efforts of visionaries like Getting and Easton, and the spirit of compromise under Parkinson's leadership was what made this revolution in precision a reality. The DoD had finally made the conceptual transition from navigation, which the Navy claimed they have been doing for centuries, to positioning. Knowing the position of all stationary and moving objects on the globe at all times and knowing the synchronized time with unprecedented accuracy opened infinite possibilities for military and civilian operations. This was indeed a revolution of the same magnitude as the initial development of the chronometer in the eighteenth century.

Because potential enemies might use GPS positioning against the United States or her allies, the civil signal was intentionally degraded through a process known as selective availability. This process reduced accuracy for civilian users and remained part of GPS as a holdover from its original military history. Selective availability was generally active, although ironically it was turned off during several national emergencies and international military campaigns when the military use of civilian receivers was widespread. It slowly became apparent that the proliferation of differential corrections in the form of augmentations rendered these perturbations totally ineffective. As a result, a Presidential Decision Memorandum was signed in 1996

which ordered the military to discontinue its use, pending justification from the DoD. In early 2000, selective availability was removed from the signals of all orbiting satellites.

During the first 25 years of GPS, several generations of satellite designs have been developed or are under development. These include Blocks I, II, IIA, IIR, IIRM, and IIF. In addition, plans for an upgraded version of GPS, known as GPS III, are currently being defined. The Block IIRM and Block IIF satellites add additional civil GPS signals at other microwave-band frequencies which should materially improve the accuracy and robustness of the civil service.

GPS CONCEPT OF OPERATION

The design objectives of the GPS system were to provide a continuously available, worldwide, all-weather, three-dimensional precision positioning system for both military and civilian users on land, at sea, or in the air, even in space. The GPS system had to operate on an accelerating platform, such as a maneuvering aircraft or missile. Additionally, the system had to be passive, or one-way, so that it could service an unlimited number of users. As a military system, the signal is required to be both jam-resistant and antispoof.

Each of these requirements drives a certain set of constraints. To be worldwide and continuously available, only a satellite system can provide global coverage, especially over oceans and polar regions. As a satellite system, frequencies less than 1 MHz skip off the ionosphere, and frequencies higher than 10 GHz are very heavily attenuated by atmospheric moisture. Satellite signal frequency was a compromise among accuracy (ionospheric delay), attenuation, and the power to be received by an omnidirectional user antenna. Thus, the selected signal was placed within the L-band for best performance. Two additional constraints were established by the military: that the satellites could be totally serviced from the continental United States (CONUS) and that the constellation could be tested by using a small number of satellites to minimize project risk. These constraints led to satellites in medium-altitude orbit. The quantitative requirements of the original GPS design were to guide a bomb to within a 10-meter circle anywhere on the planet, and to build an inexpensive (less than $10,000 at that time) device that could navigate.

GPS functions as a multilateration system: the range from at least three known locations is determined, and the point at which the three spheres intersect defines the user's location. In GPS, the system is complicated by the fact that the transmitters are moving and that the range cannot be measured directly. As a simplification, assume that the GPS satellites are stationary and that the user is upon a flat, nonrotating Earth. All of the satellites are synchronized and transmit a signal at the exact same time.

The user will receive the signal from each satellite at a different time due to the time of flight of the signal, traveling at the speed of light, across the different ranges from each satellite to the user. If the user possessed a very accurate clock that was time-synchronized with the satellites, the product of the time of flight and the speed of light would be the true range to the satellites. However, because the user is unlikely to have an atomic clock (a requirement that would make the receivers far too expensive),

the user is not synchronized to GPS time. Thus, the measured range is offset by a consistent bias and is thus referred to as pseudorange. This measurement is taken simultaneously for each satellite. Even without knowledge of the exact time, the consistent solution for the ranges based on user position and unique time bias can be computed. The satellite locations are known from the navigation message on the signal. There are four unknowns — three components of position, and time — thus measurements from a minimum of four satellites are required to solve the four simultaneous equations.

GPS SPACE SEGMENT

The space segment of GPS is the satellite constellation (see Figure 3) that consists of twenty-four or more vehicles in six orbital planes. The planes are inclined at 55 degrees and spaced 60 degrees apart. There are four satellites in each of the orbital planes, but they are not evenly spaced (this to minimize the impact of any single satellite failure). Additionally, there are typically on-orbit spares in some of the six planes. The satellites are in a medium Earth orbit at a radius of 26,561.75 kilometers (a mean equatorial altitude of 20,163 kilometers). The orbits are almost perfectly circular. The orbital period is 12 hours of mean sidereal time (a mean sidereal day is the rotation of Earth to the same position with respect to inertial space, in contrast to a solar day, and is approximately four minutes shorter than a solar day). Thus, each GPS satellite repeats the same ground track, but passes the same location four minutes earlier each (solar) day. Some of the orbital planes may have extra satellites as on-orbit spares.

The GPS payload consists of redundant

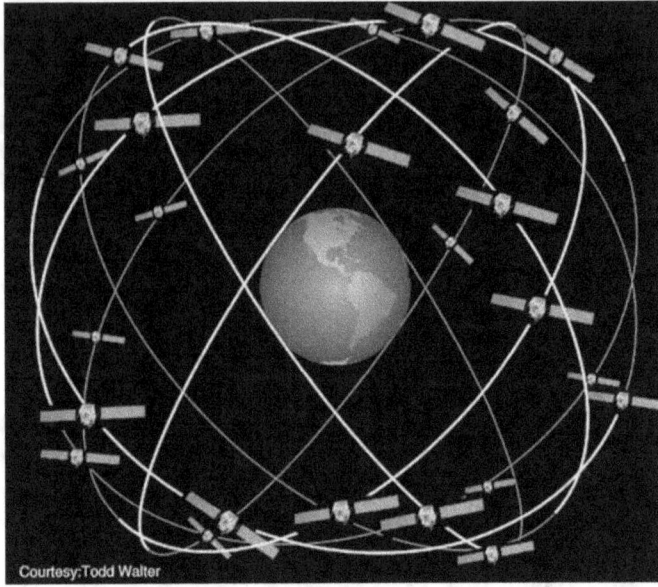

Figure 3. The GPS constellation consists of 24 satellites in six orbital planes.

atomic clocks, telemetry and control sections, and the signal-generation subsystem. The atomic clocks are rubidium and/or cesium standards that typically have long-term stability of 1 part in 1,013 per day (a drift of roughly 9 nanoseconds per day). The master control station monitors the atomic clock drift rates and models them. GPS is the first operational system known to require a correction for relativistic effects. All of these parameters are sent in the navigation message.

The satellites' telemetry subsections are responsible for receiving the uploaded navigation data from the master control station. The data is encrypted before upload to ensure than no spoofing can occur. Internal status and health are also monitored and relayed back to the control station. Note that these power levels are well below the ambient noise level. From the satellites' locations, Earth subtends an angle of approximately 14 degrees. A user at the limb of Earth is significantly farther

away than one directly under the satellites. To compensate for this greater "space loss," the antenna gain pattern on the GPS satellites is such that approximately 2.1 decibels more gain is at the edge than at the bore sight of the beam. The beam is also slightly wider than the 14 degrees of Earth to allow satellites on the other side of Earth to use GPS for positioning.

The specification for both the C/A code (coarse acquisition code) and the P/Y code (precise military code) is such that the minimum broadcast power is well below the noise floor of the in-band radiation. Using the correlation properties of the PRN codes, a GPS receiver can reconstruct the phase of the signal and use this for position and temporal information.

GROUND CONTROL SEGMENT

The GPS control segment consists of six or more monitoring stations around Earth, a master control station, and upload ground-antenna stations. Each of the monitoring stations has a set of accurate atomic clocks and tracks both the code and carrier of each GPS satellite as it traverses overhead from horizon to horizon. The monitoring stations operate at L1 and L2 frequencies to permit removing excess ionospheric delay. They also monitor atmospheric parameters such as temperature, atmospheric pressure, and humidity to permit estimating the tropospheric delay. By tracking the L-band carriers from horizon to horizon to a small fraction of a cycle, a series of 15-minute averages is created and sent to the master control station.

The master control station receives the monitoring station tracking and ground-antenna telemetry information and computes the current and predicted satellite clock offsets and satellite positions. It then converts this data to the navigation data formats described later. These rather complex satellite orbit/time filter estimating algorithms must also model the satellite solar radiation pressure; atmospheric drag on the satellite; Sun/Moon gravitational effects, including solid Earth and ocean tides; and Earth's geopotential model. Improved GPS satellite-to-satellite cross-link ranging data may also be used in the future. The navigation data is uploaded from several 10-meter diameter, S-band ground-antenna upload stations.

The navigation data is encoded on the L1 C/A signal. This data message is transmitted at the rate of 50 bps and consists of a set of 6-second subframes (ten 30-bit words) and 30-second frames. The data encoded includes the full ephemeris required to calculate the current satellite position; the satellite clock quadratic polynomial model and corrections to GPS time; almanac data used to position all the other satellites; and a handover word for P/Y-code users. The almanac data allows a user to compute the rough positions of the satellite and thus narrow the search space both in terms of PRN codes and Doppler bins.

USER SEGMENT

The user segment, or the GPS receiver, is a very sophisticated digital signal tracking device that allows converting the faint signals from the GPS satellites into an accurate position solution. The GPS receiver must process the almanac (either stored or newly acquired) to generate a search space in terms of PRN codes and Doppler frequency bins. The incoming radio-frequency (RF) signal must be amplified, down-converted

through an intermediate frequency (using a mixing process), and sampled into the digital domain. The PRN codes are correlated against the incoming digitized stream, and usually a delay lock-loop is implemented to keep the signal locked.

Once the signals are tracked, corrections are applied to the raw pseudoranges, and the position and time bias are computed through an iterated least-squares calculation. The positions are now reconverted to a useful coordinate frame such as latitude, longitude, and altitude. The original GPS "manpack" receivers were backpack-sized devices costing more than $50,000. They did, however, satisfy the original mandate to produce an inexpensive device that could navigate.

GPS has benefited greatly from the semiconductor revolution, as has the typical consumer. Civilian use of commercial GPS receivers has grown significantly since the end of the first Gulf War. A modern GPS receiver costs as little as $100 and is small enough to be embedded into a wristwatch. Additionally, the computer that calculates the position solution can support many additional features such as map displays and waypoint guidance, at some additional cost. (Figure 4.)

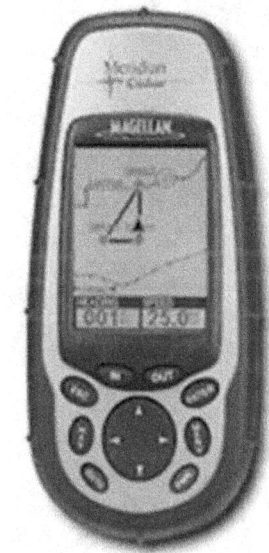

Figure 4. The Magellan Meridian GPS Receiver pictured above is typical of the type of receiver available to the public. It is a hand-held model that stores up to 20 routes and has a 9MB preloaded North American database.

DIFFERENTIAL GPS

One technique used to augment GPS is known as differential. The basic idea is to locate one or more reference GPS receivers at known locations in users' vicinities and calibrate ranging errors as they occur. These errors are transmitted to users in near real time. The errors (or their negatives, which are corrections) are highly correlated across tens of kilometers and across many minutes. Use of such corrections can greatly improve accuracy and integrity. Several large-scale differential networks have been deployed in the United States and elsewhere.

The U.S. Coast Guard (USCG) within the United States and the International Association of Lighthouse Authorities have deployed a marine-beacon differential system internationally. In the United States the system is known as National Differential GPS. The Army Corps of Engineers is currently deploying additional National Differential GPS compatible beacons to cover the entire CONUS. The Federal Aviation Administration (FAA) is currently deploying the Wide Area Augmentation System. This system is intended to provide en route navigation and nonprecision approaches for aviation users.

The FAA is also developing a Local Area Augmentation System for Category I, II, and III precision landing capability at airports (Figure 5). This will require local ground monitoring stations to ensure the integrity of the system in addition to the nominal reference receivers. The exacting requirements of Category II and III landings mandate that the Local Area Augmentation System perform many crosschecks of the GPS system to ensure integrity. If one of these

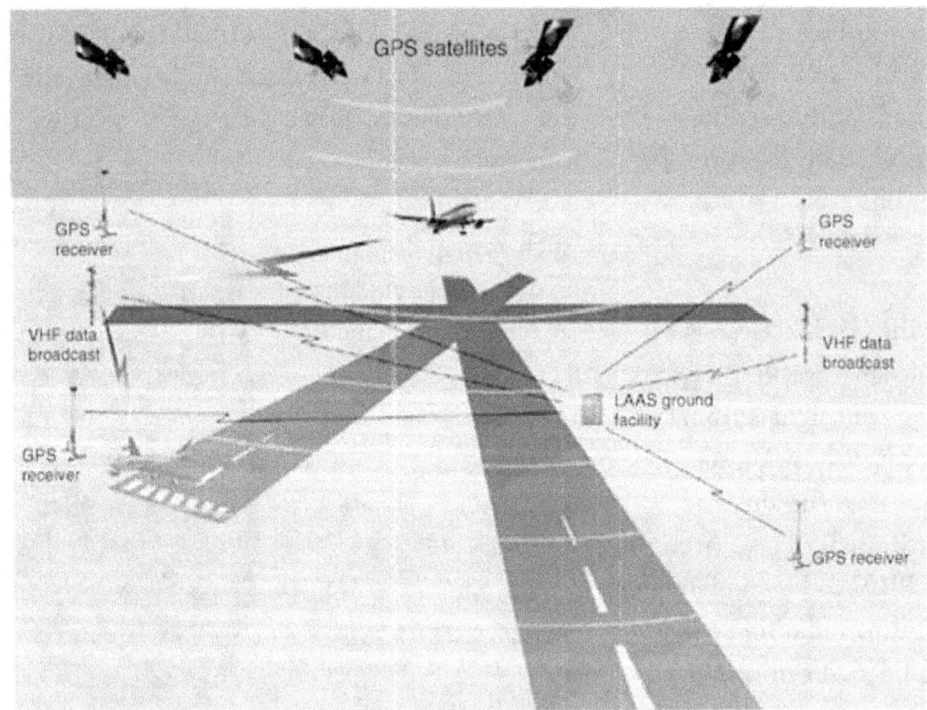

Figure 5. The LAAS system will provide precision approach capability using GPS.

crosschecks fails, the time to alarm of the local system is specified at less than 6 seconds.

Differential carrier tracking is another GPS technique that has been used by surveyors since the mid-1980s. By reconstructing the L-band RF carrier signal, a GPS receiver can attain tracking precision of 1 to 10 millimeters. Several techniques exist for resolving the integer-cycle ambiguity. Satellite motion can be exploited to do this differentially. This technique is referred to as real-time kinematic GPS. When applied, this technique provides survey-level differential positioning with relative accuracy in millimeters. Thus one can locate very rapidly an unknown point on the ground relative to a survey mark and then maintain this accuracy as the user's receiver is moved. This is now being exploited for both construction survey and real-time, automatic, machine control.

Many additional private and international systems are under development or deployed. Various private companies sell their own proprietary carrier-phase differential GPS systems for use in such diverse areas as construction, surveying, and archeology. Commercial wide-area corrections are carried by at least one commercial C-band satellite broadcast, and several oil companies have put their own differential stations on oil drilling platforms to ensure accurate positions for the helicopters and ships that service them.

MILITARY APPLICATIONS

GPS has played a role in military operations for more than a decade now. Operation Desert Storm was one of the first major military campaigns that used GPS applications extensively. Most recently, GPS played a

pivotal role during Operation Iraqi Freedom (OIF). About 70 percent of the munitions employed in OIF were precision-guided, the vast majority being aided by GPS. More than 7,000 GPS-aided munitions were reportedly used in OIF, including the low-cost, highly effective Joint Direct Attack Munition (JDAM). This contrasts with Operation Desert Storm, where only approximately 10 percent of the munitions were precision-guided.

The high levels of accuracy achievable using GPS, along with the system's all-weather capability, help explain why GPS-aided munitions are so widely used in air campaigns. During OIF, the GPS Precise Positioning Service (the more accurate signal available to "authorized" DoD users) provided an average position accuracy of 3.69 meters (compared with 24.3 meters during Desert Storm). This level of accuracy has not been seen in any previous major U.S. military engagements.

JDAMs were used in a variety of roles. The munition's GPS-aided precision allowed air strikes on leadership and command and control targets in downtown Baghdad. In fact, the 2,000-pound munitions used by F117s in the opening night leadership decapitation attempt were GPS-guided. Delayed fuzing and the uncanny accuracy of the weapon mitigated collateral damage concerns normally associated with attacking military targets in urban locations.

GPS was also an invaluable navigation aid, as fighters, bombers, and intelligence, surveillance, and reconnaissance aircraft rendezvoused in-flight with air refueling assets. Navigation aids normally available did not exist in Iraq's air control structure when combat operations began.

Additionally, adverse weather conditions and the country's featureless landscape complicated an already difficult problem as hundreds of aircraft flew to predetermined rendezvous locations. An onboard GPS-aided, in-flight navigation system was the only "nav aid" pilots and crews could rely upon to navigate safely and to arrive on-time at the correct preplanned refueling location. From there the pilot and crew would establish radar/visual contact with the tanker and effect the rendezvous.

Assistance to Special Operations Forces was often dependent upon the reliability and precision of GPS-aided munitions. In the opening days of the conflict, Special Operations Forces working with indigenous forces in the north relied on air strikes provided by B1 aircraft from the south. The B1s, accompanied by F16CJs, were routinely pulled north in the dark of night to provide air support for small teams in need of immediate assistance. The weapon of choice was JDAM because weather conditions often precluded the use of laser-guided munitions and bomb-laden fighters were unable to provide meaningful on-station time due to the distances involved. Uncontrolled strikes were sometimes a reality when an airborne B1 was unable to establish voice contact with a tactical air controller on the ground. Ground forces insisted on support despite their inability to establish contact. Target coordinates were relayed through a Special Operations Forces control element to the combat air operations center, passed to an airborne E3, and then communicated to the on-site B1. Despite their inability to see the target, B1 crews were successful in every attack because of the accuracy and precision

of their 2,000-pound GPS-aided munitions.

Russian-built Iraqi jammers detected on the first day of OIF were taken out by a U.S. military trained to deal with this new threat. The 527th Space Aggressor Squadron was called in to advise the Combined Forces Air Component Commander on the evolving GPS-jamming situation in Iraq, and it suggested courses of action. Ultimately, the jammers were destroyed with GPS-aided munitions from F117 and B1 aircraft.

The bottom line is that OIF marks an important milestone in Space history. For the first time, an adversary attempted to deny the United States its access to Space. This attempt was completely ineffective due to the advanced preparation for exactly this type of situation. The Aggressors of the 527th were instrumental in helping our forces train like they would have to fight.

SELECTED APPLICATIONS

Applications of GPS have continued to multiply, as commercial and civil organizations apply creativity in using its capability. This section will not attempt to enumerate all current and future potential uses. Instead, selected examples will illustrate the revolutionary advances that have been made possible by this remarkable system. Many of the topics presented are at the cutting edge of current research and may profoundly improve our understanding of our world as well as improve our productivity and safety.

Survey and Crustal Motion

Until the advent of carrier-phase differential GPS, measuring the relative distance or motion of large objects accurately over time required painstaking surveys using laser interferometry, and they tended to be one-dimensional. However, carrier-phase differential GPS that can track 3-D relative positions down to millimeter levels across very long distances is revolutionizing the field of geomatics. Currently, experiments are underway that monitor the relative positions of the mountain sides of several volcanoes in the states of Hawaii and Washington. Previous attempts to perform these kinds of experiments proved difficult due to the requirement for consistent line-of-sight measurements using optical sensors. Data recorded by using survey-quality GPS receivers have detected bulging of the mountains and are providing insights that may one day enable scientists to predict volcanic eruptions.

Similarly, hundreds of GPS receivers have been placed along fault lines throughout California and other parts of the world to validate theories about plate motion and gain valuable information on preconditions for earthquakes. Again, research in this area is still in its infancy, but never before has it been so economical or in some cases even possible to measure distance across large geographic features down to the millimeter level. At this time, data is being gathered to validate crustal motion that will certainly lead to refinements in these models.

Aviation

The aviation industry has been an early adopter of GPS technologies, and it remains at the forefront of developing and implementing advances in GPS. In the early 1990s, a prototype GPS landing system for Category III (zero feet ceiling, zero miles visibility) was developed and demonstrated by Stanford University under an FAA grant. This system used carrier-phase differential GPS to ensure a correct position. To resolve the integer-cycle ambiguities quickly and robustly, two ground transmitters that broadcast GPS-like signals were used to augment the system. These "pseudolites" exhibited a large change in Doppler shift due to the rapid geometric change. The resulting system demonstrated more than 100 auto-coupled landings at Crows Landing Airport in California; data was independently validated by using the Crows Landing laser tracker. The data showed an accuracy of better than 0.5 meter (3-D) in the final phase of landing.

During one of the auto-coupled landings, a satellite upload from the Master Control Station caused the satellite to interrupt its transmission for approximately 1 millisecond. The Stanford system detected this glitch in the space segment and called off the landing in real time.

Recently the FAA has approved the GPS as a precision navigation aid, but most general aviation and commercial pilots have been using GPS as a backup system for years. Additionally, modern aviation GPS units are programmed with a full aviation database and can notify the user of airspace violations. In an emergency, these units can guide the pilot to the closest airport at the touch of a button.

GPS, as a full thirteen-state sensor for an aircraft, provides a powerful suite of information at a relatively low cost. Combined with inexpensive computer graphics, a synthetic out-the-window perspective display can be used to improve vastly the presentation of critical data to the pilot. The futuristic vision of tunnelsin-the-sky for improved navigation is being tested today in various laboratories around the world. Pilots who have experimented with these systems report a much reduced workload and greater situational awareness. The potential to reduce controlled flight into terrain could save many lives currently lost due to such accidents. Likewise, if all other aircraft are prominently displayed, midair collisions can be reduced. These displays have also shown great promise in enabling closely spaced parallel approaches in inclement weather. This alone can save the United States billions of dollars in runway expansions and avoid environmental impacts that such construction would have on surrounding areas.

Vehicle Tracking

The so-called "urban canyon" can adversely affect GPS, but vehicle tracking remains a very important application. During urban-canyon outages, most vehicle-tracking systems use some form of inertial augmentation to provide a position solution. Very small nano-technology inertial units of postage stamp size will make this solution even better and cheaper. Commercial companies have great interest in knowing where their equipment is currently located, and GPS provides an ideal answer. Many cities now have buses equipped with GPS receivers and radio transmitters. Each bus stop has a display of the current location of the next bus and an estimate of the time until it arrives. Likewise, many cities have GPS equipment on their emergency service vehicles to better manage their response. This has been shown very effective in reducing response time and managing scarce resources during a large-scale disaster.

Vehicle tracking yields a great competitive advantage to a corporation. In one case, a cement company in Guadalajara, Mexico, would send fully loaded cement trucks into the city every morning, even though orders had not yet been placed. Using simple radio communication, this company responded to orders in less than half the time of any of its competitors. Though several trucks would return without having had a customer by the end of the day, within a short time this company dominated the cement delivery market.

Law enforcement officials have been able to use GPS to remotely monitor suspects and increase their effective manpower. Obtaining a court order allowing them to install a GPS receiver surreptitiously on a suspect's car, Seattle police were able to reconstruct the time and path of a suspect's location during a two-week period, without alerting the individual to the surveillance. The information they obtained led directly to evidence that convicted the suspect.

Precision Munitions

No discussion of GPS would be complete without a discussion of military applications. In spite of its extensive use in many civil applications, GPS was designed primarily as a military system, and to continually develop, GPS must fulfill its primary mission. Several military applications for GPS were developed in recent years. An example is the JDAM. This precision-guided munition has demonstrated a battlefield accuracy of better than 10 meters. The trend in the future is to reduce the size of the explosive warhead on these munitions, and this can be achieved only if the guidance system is capable of pinpoint accuracy. (See a broader treatment in a companion paper at this symposium.)

In an example of a purely defensive military application, the DoD recently deployed a Combat Survivor/Evader Locator radio for service members. This radio allows downed pilots to relay their positions to rescuers directly to enable rapid rescue and minimal exposure to hostile forces. The Combat Survivor/Evader Locator replaces four different individual devices with a single integrated package.

Space Applications

Some of the most innovative and unusual applications of GPS occur in the area of Earth-sensing and space applications. Low Earth-orbiting satellites can use GPS to measure both position and attitude. Precise satellite data can be used to refine gravitational models of Earth and as a sensor for attitude control. A soon-to-fly satellite experiment, the Gravity Probe B, uses very precise spherical gyroscopes to yield a quantitative measurement of Einstein's theory of relativity. For the experiment to be valid, Gravity Probe B needs to fly a drag-free polar orbit to within 100 meters. GPS is used to provide guidance information to position the orbit of the satellite initially. One of the most unusual applications of GPS is using the reflection of GPS signals from waves at sea to detect wave height in the open ocean.

FUTURE IMPROVEMENTS

The first Block IIR GPS satellite was launched in 1997. Though later versions of Block IIs will be a bridge to a future GPS system, known as GPS III, the next generation of GPS is still being defined. Future improvements in the GPS system are driven by competing civil and military requirements. All users desire more signal power to ensure resistance to interference and/or jamming. In the last decade, GPS has become essential to virtually all DoD operations. International constraints on RF spectrum availability dictate that improvements remain within the radio navigation bands. On the civil side, the expectation has become that GPS will remain continuously available across the globe for the foreseeable future. Civilian users are urgently requesting the second and third frequencies to calibrate ionospheric delays and provide a backup if the L1 signal is jammed.

Several key advances are planned for the end of the Block II series of satellites. The most important are two additional signals on the Block IIRMs and three on the Block IIFs. The first additional signal is a replica of the C/A code but at the L2 frequency. This will allow direct measurement of ionospheric errors for civilian users. Military users will have a new split-spectrum code (called M-code) on both L1 and L2. This code has the advantage of transmitting most of its power in the nulls of the C/A code, maximizing spectral separation.

The Block IIFs will include yet another civil signal at L5 (1,176 MHz). This signal is intended to be a higher accuracy signal, which implies a higher chipping rate and a longer code sequence. Likely, it will include an unmodulated channel to enable much longer integration time for superior noise rejection. Other technical advances for the late IIFs include intersatellite communication as well as improvements in the onboard rubidium/cesium clocks. Likewise, upgrades in the ground station facilities will reduce the errors in ephemeris predictions. For GPS III, the need for further increases in M-code power will probably lead to a spot beam of about 1,000 kilometers.

Though all specifics of the GPS III concept are still to be determined, as well as the details of the relationship between the U.S. GPS and the European Galileo, the United States intends to continue providing and improving on a worldwide, continuously available, precise navigation signal free to all. GPS III will undoubtedly follow that tradition and provide a yet more robust and more accurate system of positioning on a global scale. The one thing that is most certain for the future of GPS is that its uses will continue to multiply and impact our daily lives.

Relationships to Galileo

Galileo is the European version of GPS. The European Union is committed to building a thirty-satellite, civil space-based navigation system at an estimated cost of 3.4 billion euros. The initial funding of 547 million euros is intended to fund the study and development phase, which is expected to take approximately three years. Galileo will be an entirely civil system that promises to be independent but interoperable with the civil components of GPS.

Several outstanding issues must be resolved before Galileo becomes operational (planned for 2008). The most crucial is that the Galileo signals not interfere with any of the GPS signals. Ideally, Galileo would use a compatible geodetic reference frame and time base calibrated to GPS. This would present the Galileo satellites as an augmentation to the GPS constellation or, conversely, the GPS constellation as an augmentation to Galileo.

Barring this level of interoperability, it is likely that Galileo's time base and geodetic reference frame will be distinct from GPS's but will be easily translatable if real-time data is available. The exact configuration of the Galileo system is not yet certain and is the subject of current diplomatic negotiation between the United States and the European Union.

CONCLUSION

In the last ten years, GPS has become an international utility with more than twenty million users. Recent conflicts have conclusively shown it to be essential to precision warfare. Especially important is the stringent limits on "wild weapons" that can lead to undesired collateral damage. Thus the U.S. military depends on GPS for virtually all combat operations.

Already many civil developers are working on augmentations to cell phones that will include GPS receivers. This promises to increase civil use to hundreds of millions of users. The advances in navigational/location science brought about by GPS are also significant to future space exploration. However, worldwide dependency on this global utility requires an increasingly robust system. With the advent of Galileo, the roughly fifty-five operational navigation satellites should provide the needed robustness and lead to even further innovative uses.

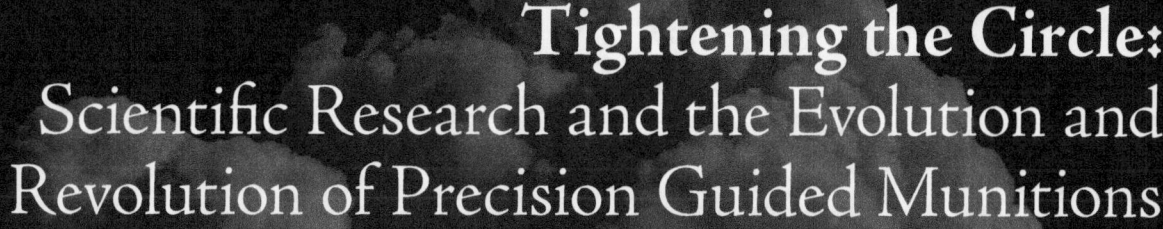

Tightening the Circle:
Scientific Research and the Evolution and Revolution of Precision Guided Munitions

Lieutenant General Daniel P. Leaf, USAF

Robert I. Sierakowski

Robert P. White

Abstract

Weapons designed to strike the enemy at ever-increasing distances and with ever-increasing precision and lethality are the result of critical scientific discoveries funded by the Department of Defense. The development and refinement of Precision Guided Munitions (PGMs) can be traced to initiatives taken before and during World War II to increase the effectiveness of aircraft-delivered ordinance.

The real revolution in PGMs occurred in the 1970s with the development and maturity of laser research and space-based initiatives such as the Global Positioning System. Beginning with their use in Vietnam, through the 1991 Gulf War, Kosovo, and Afghanistan, and through their latest employment in Iraq, the efficacy of PGMs has improved markedly, as evidenced by the incremental improvements in lessons learned.

These improvements will continue due in large part to the technologists and the warfighters who, in concert with military and civilian visionaries, conceived unique concepts and possibilities with the marriage of seemingly disparate discoveries. The results have been revolutionary in the art of warfare.

INTRODUCTION

In 212 BC Archimedes, the world's first warrior-scientist, employed scientific research and applied the resulting principles to his war machines to fight the Roman invasion of his Greek city-state of Syracuse. The weapons he employed kept the enemy at bay and effectively struck the Romans from a distance while ensuring the safety of the Syracusan warriors.[1] Archimedes' system of catapults, levered arms, and his fabled mirrors that focused the rays of the sun against the Roman galleons were ancient conceptual forerunners of the sophisticated precision guided arms employed today (Figures 1).

Figure 1. Archimedes' system of catapults, levered arms, and his fabled mirrors that focused the rays of the sun against the Roman galleons, were ancient conceptual forerunners of the sophisticated precision guided arms employed today.

Man's search for sophisticated weapons of war, particularly weapons that could be delivered from ever-increasing distances with accuracy and telling effect, has been an elusive goal until relatively recently. The first half of the twentieth century witnessed two world wars and several regional conflicts when hundreds, if not thousands, of bombs were required to destroy selected ground targets. Often, these bombing attempts were wholly unsuccessful. Attacks with unguided iron "dumb" bombs was termed area bombing, and not without good reason. Indeed, entire cities were considered appropriate target areas. During such missions, large numbers of lumbering aircraft and many crewmen were put at risk, and many were lost. The requirement to destroy targets effectively with a minimum of force and with minimum threat to the attacking forces was obvious.

The search for weapons designed to strike an enemy at ever-increasing distances and with ever-increasing precision and lethality while ensuring the safety of the attackers depended on key scientific discoveries and an organizational structure capable of taking advantage of them. In the latter part of the twentieth century, the vast majority of these critical discoveries were funded by Department of Defense (DoD)–sponsored research programs.[2] Not until the late 1960s, though, did scientific advances allow the marriage of various technologies to permit the development of what are now termed precision guided munitions (PGMs). As it stands today, we have indeed developed an impressive PGM capability, but it need not have taken as long as it did. Owing to historical vagaries of research and development (R&D) funding in munitions and delivery systems, precision munition development sputtered along until discoveries made in the past joined with contemporary realizations to produce yet another "eureka moment." What is significant about the PGM developmental process is that most of the applicable technology was available for quite some time. Even before the Wright Brothers first flew, precision-guided technology was "in the air."

The Wrights may have been the first to take to the air in manned flight at Kitty Hawk in December 1903, but in September of that year at an exhibition in Germany, Karl Jatho helped set the stage for further advances in unmanned airborne weaponry. Jatho designed and flew a pilotless biplane 12 feet long powered by a 9.5 horsepower gasoline engine for a longer sortie than the one the Wrights would soon attain. In 1908 three Americans, Dr. Elmer Sperry, Dr. Peter Cooper, and Lieutenant Patrick Bellinger, wed this pilotless capability to another technological breakthrough with the invention of the automatic gyroscopic stabilizer, which permitted an airborne craft to fly straight and level.[3] With the blossoming of powered flight, both manned and unmanned, the prospect of bombing from the air was imminent. In 1910, Glenn Curtiss demonstrated the concept of aerial bombing using simulated bombs on a dummy battleship. U.S. Army tests using live bombs quickly followed — this, ten years before Billy Mitchell sank the *Ostfriesland* off the Virginia Capes![4] In 1911, eight years after the Wrights first flew, aircraft bombardment was first used operationally when Italian aviators dropped bombs on Turkish forces in the war over Libya.[5]

In 1916 the European powers had been at war for more than two years, and the U.S. War Department recognized it needed a developmental, testing, and procurement facility for aviation needs. It established an experimental station and proving ground at Langley Field, Virginia. While the Army's Air Service was getting up to speed in its R&D armament program, the U.S. Navy in 1917 actually conducted tests to take the pilot and bombardier out of the bombing loop. The Navy funded the first radio-controlled aerial torpedo, a converted Navy Curtiss N–9 trainer powered by a 40-horsepower engine capable of flying 50 miles with a 300-pound bomb load.[6] The Army Air Service soon pulled ahead with a much more sophisticated unmanned aircraft designed and tested by Charles F. Kettering, an automobile engineer with Delco Corporation, the predecessor of General Motors. At the time, Colonel Henry (Hap) Arnold, arranged for the field testing in France of what became known as the Kettering Bug, the first unmanned aerial vehicle (UAV) to be mass produced.[7] Having a maximum range of 100 miles, a Sperry gyroscope held the "Bug" on course. An aneroid barometer maintained its altitude until the craft shed its wings after a predetermined number of propeller rotations, where upon it plunged to earth, and its 300-pound warhead, loaded with mustard gas and high explosives, would detonate.[8] Exactly where it fell to earth was quite problematic. The problem would have to wait for another war, for neither the Bug nor the Curtiss N–9 was operationally employed during the conflict. What became of America's initial endeavor with its first guided bomb, the Kettering Bug? According to "An Automatic Bomber" that appeared in the 24 May 1919

Figure 2. Sperry Aerial Torpedo, first flight 1918
Direct ancestor of the modern cruise missile.

Figure 3. Kettering "Bug"
The first mass produced guided weapon.

issue of the *Army and Navy Journal*, the Bug was "placed in the secret archives of the War Department at Washington, there to remain, it is hoped by the inventor, for all times."[9]

The aerial bombardment weapons that actually did see operational use during World War I were quite crude. Early bomb designs consisted of rejected artillery rounds modified to accept stabilization fins, and most bombs weighed no more than 100 pounds. Aiming was primitive. In employing these new weapons, belligerents soon realized that hitting where you aimed was a significant challenge.[10] Many variables affected targeting. The ground target might or might not be stationary, but the aircraft was in motion. The aircraft's speed, direction, and acceleration combined with the direction and speed of the wind plus the air temperature and humidity all affected the fall of its bomb load.[11] Bomber aircraft with their external carriage-mounted ordnance struck what they considered to be strategic targets, but by today's standard they would be tactical targets, and the bombers of World War I, just like the bomber aircraft to come in World War II, were used in mass formations to compensate for relatively small munitions and lack of precision. In addition, pursuit planes, or fighter escorts, were a requirement to safeguard the low and slow lumbering bombers.[12]

WORLD WAR II

By the end of World War I, American Air Service forces, with only 150 strategic (interdiction) bombing raids to their credit, had obtained relatively little experience in this new endeavor.[13] With little bombing experience, there was no new doctrine to provide the rationale to acquire new technology. To the Air Services' credit, a dedicated R&D center McCook Field, Ohio, provided a facility where the Army conducted its wartime aeronautical engineering initiatives that continued throughout the interwar period. While advances in individual airframe technology, bombsights, and doctrine were obtained, little thought was given to munitions technology, and no War Department funding was available for any type of guided air weapon.[14] Those who, like Douhet, conjectured about getting bombs on

Figure 4. Bomb technology changed little between World War I and World War II.

Figure 5. B–17 mission during "Big Week," 20–25 February 1944. Area bombing was a hit or miss proposition that was costly in men and machines. During "Big Week," 3,300 bombers were dispatched from England and 500 from Italy, with 137 of the former and 89 of the latter being lost. Also, 28 Army Air Forces fighters were shot down. The number of U.S. personnel killed, missing, and seriously wounded totaled 2,600.

Figure 6. Ploesti, Romania, 1 August 1944. Low level attack ensured bomb accuracy—but at a huge cost. Although overall damage to target was heavy, of 177 planes and 1,726 men who took off on the mission, 54 planes and 532 men failed to return.

target made highly erroneous assumptions: that vital targets could be identified and found, that they could then be hit from altitude, and that they could be destroyed.[15]

As we were to learn in World War II, a B–17's fire control did not have pickle-barrel parameters. Except for the fact that the bodies of the bombs were redesigned to make them more compact for internal bomb-bay storage, the bombs used in World War II had changed little since the previous war, and little thought had been given to their fire control system.[16] The B–17's fire control system, the super-secret Norden bombsight, began as a U.S. Navy program to design a precision bombsight to target maneuvering enemy ships, but the Navy decided instead to use dive bombing to achieve that end.[17] Because U.S. Army Air Forces

bombers could not dive-bomb their targets, they employed the Norden bombsight to increase their chances of scoring direct hits. In one significant example from July 1944, forty-seven B–29s raided the Yawata steel works in

Japan from bases in China and dropped 376 general-purpose 500-pound bombs. One bomb hit the target, which represented 0.25 percent of the bombs dropped during the raid![18] In the European theater, things weren't much better: In the fall of 1944, only 7 percent of all bombs dropped by the Eighth Air Force hit within 1,000 feet of their aim point.[19]

Hit Probability of 90% against a 60' x 100' Target with the Use of 2,000-Pound Unguided Bombs at Medium Altitude[20]

War	Number of Aircraft	Number of Bombs	CEP (in feet)
WW II	9,070	3,024	3,300

Given that World War II bombing accuracy proved to be a contradiction in terms, it seems an effort would have been to remedy the situation. When Hap Arnold became the commander of the Army Air Corps in September 1938, he immediately accelerated Air Corps R&D efforts across a wide spectrum of long-term technologies and committed the funds for massive new R&D facilities at Wright Field, Ohio.[21] He did this because he realized that in the interwar period relatively little thought and investment were being directed toward futuristic weapon systems. His assessment was confirmed when he read the results from a 1939 special air board that he had commissioned to look at future weapons requirements: the board made no mention of jet propulsion or missiles, much less true precision bombing.[22] Arnold realized that the National Advisory Committee for Aeronautics, the arbiter of aircraft design and innovation since its creation in 1915, "had made vital contributions to aircraft hardware and design…but the organization was doing so without considering how the developments fit into the Air Corps's balanced air program."[23]

In late 1939 after witnessing the effects of *blitzkrieg*-based warfare and realizing the need to create an effective American fighting air force, Arnold drastically curtailed his basic and applied research programs and concentrated on R&D. He focused on improving current technologies, among them the B–17, B–29, rocket propulsion, and glide bombs. It is interesting to note that during the World War II years, while Arnold was committed to the Army Air Forces doctrine of precision daylight bombing (as opposed to the British Bomber Command's nighttime area bombing philosophy), he was also committed to procuring standoff and remotely controlled weapons to protect his pilots from harm.[24] As early as the summer of 1939, Arnold was in touch with Charles Kettering and discussed resurrecting the World War I–era Kettering Bug, but this idea was shelved due to range requirements. When Arnold made a visit to England in 1942, he noticed that even in cities, many bombs fell in open areas. With this thought in mind, he concluded that a bomb gliding to its target on a flat trajectory would, by default, hit a vertical surface.[25] What resulted was a requirement for a glide bomb (GB) that would glide one mile for each 1,000 feet of altitude and would have a circular error probable (CEP) of less than one-half mile.[26] The resulting weapon, the GB–1, a 2,000-pound bomb fitted with wooden wings spanning 12 feet, was targeted against the German city of Cologne in the spring of 1944.[27] Operationally, B–17s carried one GB–1 under each wing with the bombs being released about 20 miles from the target, but with an average range-miss distance from 3,000 to 5,000 feet and from

Figure 7. The GB–4 (above) was a GB–1 (AZON) with an added television camera and radio remote control, which during testing achieved an impressive 200 foot CEP.

Figure 8. The VB or vertical bomb series, (below) consisting of 1,000 pound bombs fitted with a steerable tail assembly and a tracking flare that were ultimately equipped with a radio control system, allowing a bombardier to direct the bomb to a target using a joystick.

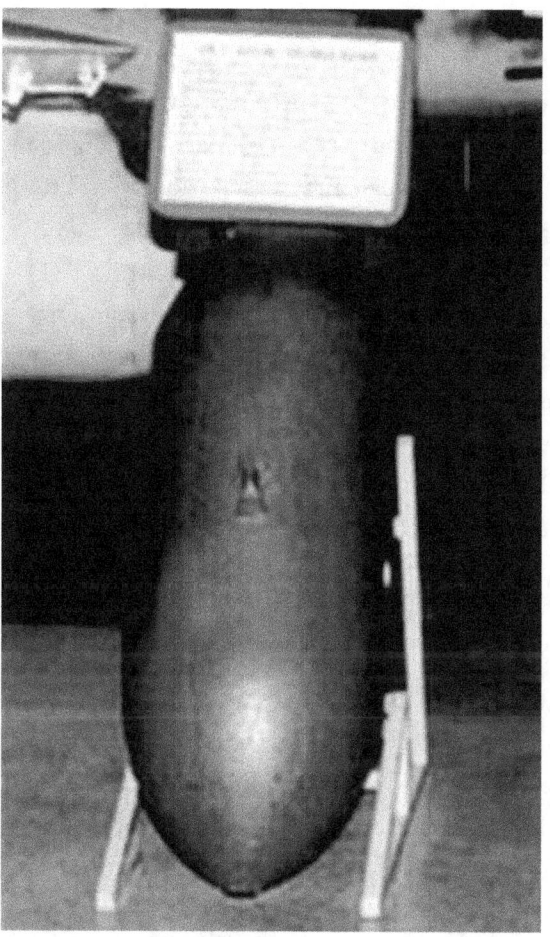

700 to 1,000 feet in azimuth, the best that could be hoped for was that the bombs would fall within the city limits, which they did.[28] To Eighth Air Force commanders, it was simply an area bombing weapon, and they rejected it. Consequently, the entire developmental glide bomb program was diminished to the point that almost all glide bomb initiatives did not see combat.[29] In addition to the targeting shortcoming, only two such weapons could be carried by each aircraft, significantly increasing

Figure 8. The VB–9 guided bomb, equipped with a radar seeker.

drag and reducing the B–17's speed and range.[30] By comparison, our allies had better results with their guided weaponry. One may even date the precision weapon era to 12 May 1943 when a Royal Air Force antisubmarine Liberator dropped an MK–24 acoustic homing torpedo that found and seriously damaged the U–456.

The damage forced the submarine to the surface, where convoy escort vessels quickly sank it.[31]

Though operationally limited, the U.S. glide bomb program was developmentally dynamic; fifteen types actually came from the Wright-Patterson weapons laboratories in Ohio. While the GB–1 was guided by a gyroscopic stabilization system, others were fitted with a variety of guidance systems including light contrast, infrared, active radar seekers, and television imagers.[32] The only other glide bomb to see combat, the GB–4, was a GB–1 with an added television camera and radio remote control that achieved an impressive 200-foot CEP during testing.[33] The poor quality of the GB–4's television technology and its inability to penetrate hardened concrete targets limited its usefulness.[34] Also coming from the Wright labs was the vertical bomb (VB) series, 1,000-pound bombs fitted with a steerable tail assembly and a tracking flare, and ultimately equipped with a radio control system that allowed a bombardier to direct the bomb to target with a joystick.[35] This azimuth only (Azon) weapon was used extensively to target bridges in Burma, the World War II chokepoints of the Japanese supply chain. Of the 459 Azons used in Burma, 27 took out bridges: a success rate of nearly 6 percent. This particular guided weapon concept was not new, for the Germans had experimented with such weapons during the First World War and continued their tests in the 1930s, ultimately refining several guided weapons that were successfully employed during the Second World War, in one instance, sinking the 42,000-ton Italian battleship, *Roma*.

KOREA AND VIETNAM

By the end of the World War II, both the U.S. Army Air Forces' and the U.S. Navy's weapons laboratories had produced relatively sophisticated guided weapons programs based largely on scientific advances made by the weapons guidance division of the National Defense Research Committee.[36] Technical advances in the field of guided weapons included radio remote control, a light contrast seeker warhead (VB–5), a heat seeker (VB–6), a series of television bombs, and a radar seeker (VB–9).[37] In fact, the guided weapons program became the Army Air Forces' third largest development program, exceeded in investment only by the unguided bomb and jet propulsion initiatives.[38] Regardless of the large degree of investment, as with all new technology, drawbacks persisted. These new guided weapons "generally required clear weather, easily identifiable targets, and air superiority" not to mention that many aircrews found the visually guided weapons delivery process to be far more dangerous than a conventional bombing mission, given the amount of time required over the target area.[39] This issue would repeat itself in the Korean and Vietnam conflicts.

In retrospect, except for the laser, the scientific and technical basis of current PGMs can be traced to initiatives taken before and during World War II to increase the effectiveness of aircraft-delivered ordnance. In the United States, these initiatives were generally met with official indifference and the consequent lack of funding.[40] Much the same can also be said for the period following World War II when emphasis shifted to the perceived requirements for the Cold War:

nuclear weapons, jet aircraft, and missiles.[41] The Azon program was canceled and the limited R&D funds available for guided weapons development for the now-independent Air Force were spent on the potentially more capable Razon, which was controllable in both the vertical and horizontal axes.[42] When war broke out in Korea in June of 1950, several Razon-modified B–29s were on-station in Guam, but so were the very much neglected, rusted, and damaged tail assemblies of the Razon-modified 1,000-pound bombs. After a rough start, the Razon-equipped bombing units took to the air and succeeded in destroying several bridges, but it took, on average, four direct hits from the thousand-pounders to bring down these relatively narrow, hard targets.[43] The Razon had potential but lacked the required punch for large reinforced targets. The result was a marriage of Razon guided-bomb technology and World War II–era British big-bomb technology: the 12,000-pound Tarzon guided bomb seemed the right answer when, in early 1951, it took out several bridges in one shot.[44] The subsequent

Figure 10. The B–29 carried the 12,000 pound TARZON (VB), used during the Korean War.

loss of one Tarzon-equipped B–29, the near loss of another, and the complex training and loading procedures required suspending the program. Even with a rather sketchy performance record, the latest guided bombs proved they could take on small, tough targets, and could do so with far better precision and with fewer sorties than conventional bomb missions could.[45]

The results of the Korean War guided-bomb experience proved little different from what was learned in World War II: clear weather and air superiority were necessary to deliver the bomb on target.[46] While conventional iron bomb delivery had improved during the Korean War, the idealized process of precision bombing remained elusive.

Hit Probability of 90% against a 60' x 100' Target with the Use of 2,000-Pound Unguided Bombs at Medium Altitude[20]

War	Number of Aircraft	Number of Bombs	CEP (in feet)
WW II	9,070	3,024	3,300
Korea	1,100	550	1,000

One historian, in commenting on the development of precision guided bombs during the 1950s, makes the point that the problems with the then-current guided weapons could have been solved with relatively little engineering effort. "None of the problems were very complex nor required new scientific discoveries for solution…they were engineering puzzles that were less dependent upon inspired imagination and more responsive to time and effort," rather than basic research.[48]

If the problems facing precision bombing were relatively easy to solve, why weren't they? The answer lies in how Air Force R&D was organized, especially when it came to something as basic as bombs. From the 1907 establishment of the Aeronautical Division under the Signal Corps, until 1962, fifteen years after the Air Force became a separate service, the U.S. Army Ordnance Department was responsible for the development of all high-explosive, fragmentation, and semi-armor-piercing bombs. Basically, weapons that departed the aircraft became the responsibility of Army Ordnance, and those that stayed with the plane belonged to the Air Service/Air Corps/Air Force. To make matters worse, incendiary bombs were the responsibility of the Army's Chemical Service, and the development of armor-piercing bombs was the responsibility of the Navy's Bureau of Ordnance. In essence, no strong air advocate existed for precision guided ordnance.[49] The seeds for such an advocate were born in December 1949 when the Air Material Armament Test Center was established at Eglin AFB, Florida, to "concentrate at one location the widely scattered activities engaged in air armament development."[50] Separate from the Air Proving Ground Command, which had been operating

at Eglin since the 1930s, this small organization was made responsible for nonnuclear air armament developmental testing in the middle of the acquisition cycle. Funding priorities after the Korean War, though, emphasized strategic nuclear warfare, and any money allotted to tactical bombing efforts went toward tactical nuclear weapons and its supporting structure. Again, precision guided weapons development for the Air Force was put on hold.[51]

The same could not be said for the Navy, which developed the 250-pound Bullpup and 1,000-pound Bulldog guided missiles as an answer to their Korean War bridge problem. The 1952 Bullpup U.S. Navy requirement reached initial operational capability in 1958 and entered Air Force stockpiles in the early 1960s, and the bigger Bulldog reached initial operational capability in 1964. Both weapons were based on the same technology as that found in the World War II Razon and Korean War–era Tarzon bombs, except for the addition of rocket propulsion.

This daytime, clear weather weapon had its

Figure 11. USAF B–29 "Superfortress" equipped with Bullpup guided missiles. In April 1965, USAF and USN Bullpups and Bulldogs "bounced off" the Thanh Hoa "Dragon's Jaw" bridge in North Vietnam.

drawbacks, and twenty years after the end of World War II it was not much better than its predecessors. Proof was forthcoming when the Bullpups and Bulldogs went up against the Thanh Hoa railroad Bridge in North Vietnam in the spring and summer of 1965, only to strike their target, explode, and, as one pilot described it, just "bounce off."[52] The Thanh Hoa Bridge was called the Dragon's Jaw by the North Vietnamese, and throughout Rolling Thunder, U.S. pilots acknowledged it as the bridge that "would never go down."[53] In the first attack against the Dragon's Jaw on 3 April 1965, 79 F–105s dropped 638 750-pound bombs and fired 298 rockets. Five aircraft were lost. The bridge was scratched but remained very serviceable.[54] Over the next seven years, U.S. airmen paid a heavy price in their many attempts to down the bridge using dumb bombs and performance-limited guided weapons based on World War II–era technology. The answer was to come in two parts: technological change, if not revolution, in the research lab was occurring and was slowly making its way into aerial weapons applications, and several bureaucratic lines were being breached.

Three significant changes occurred with respect to Air Force guided weapons in the years leading up to the Vietnam War: two were organizational, the other was technical. In 1962 Army Ordnance relinquished its 55-year-old grip on bomb development and it finally became an Air Force responsibility. Organizationally, the other change was that part of this responsibility now rested in the relatively obscure Eglin-based Detachments 4 and 5 of the Research and Technology Division of Air Force Systems Command. What had begun in 1949

as the Air Material Armament Test Center had finally found its voice, and in 1964 it argued that something should be done.[55] No one knew it at the time, but thanks to the DoD basic research funding in the 1950s that resulted in the invention of the laser, a revolution in precision bombing was at hand. With the first operational laser demonstration in 1960, new technology was now available, and Detachment 4, soon to become the Air Force Armament Technology Laboratory (AFATL) in 1966, became a strong advocate for its development.[56] Detachment 4 began its laser research in 1961 and continued its experiments through 1965. Coincidentally, the Air Staff at this time concluded that tactical conventional ordnance was abysmally inadequate and recognized the benefits inherent in standoff launch capabilities.[57] But the key requirement to make that possible came not from the Air Staff nor from the laser experiments ongoing at Detachment 4, but from U.S. Army scientists who were interested in applying the power of the laser to ground warfare.

Between 1962 and 1965, scientists at the Army's Missile Command in Huntsville, Alabama, worked on producing a pulsed laser generator as well as a laser detector that could identify reflected laser light. They envisioned a soldier using the laser to help guide antitank missiles, but the enemy threat in the early years of the Vietnam War did not include armor, so the laser program lost its urgency.[58] Thus, in 1964, laser enthusiasts at Huntsville crossed bureaucratic lines and brought their research results to the Aeronautical Systems Division at Wright-Patterson. The Systems Division gave the information to Detachment 5 at Eglin, whose charter "was to seek out ways in which technology could be employed to bring about an immediate improvement in the combat potential of air weapons in Southeast Asia."[59] So it came to be that, in the fall of 1964, the commander of Detachment 5, Colonel Joe Davis, watched as Martin Marietta Corporation engineers demonstrated in Orlando, Florida, what they called a laser.[60] As he watched the laser image remain focused while its reflected shine tracked a piece of plywood 2,000 feet distant, Colonel Davis realized, "we ought to get this laser to steer our bombs."[61] Detachment 5 launched its laser bomb development program in the spring of 1965, and by late 1967 the Air Force was ready to test the results in Vietnam with a 750-pound version. The laser-guided bomb (LGB) tests in Vietnam throughout 1968 made it clear that this new weapon would be very effective on high-value targets. By 1971, kits had been designed for 500-, 1,000-, 2,000-, and 3,000-pound bombs.[62]

As a result of the 1968 bombing halt, the North Vietnamese lines of communication and supply had enjoyed a four-year respite from attack. When the North Vietnamese launched their Easter Offensive of 1972, the new Paveway LGB was ready. The weapon was used in two significant ways: against the main ground and armored invasion thrust, and in the early May resumption of bombing in the North.[63] With regard to stopping the infantry and armored thrust, the U.S. Air Force achieved a significant success against tank and artillery forces with LGBs.[64] Battlefield chokepoints were targeted. As the Seventh Air Force commander put it, "we earmarked a certain number of F–4s on a daily basis with LGBs [and] began the destruction of these points. Such was the accuracy of a laser bomb…we were getting 6 foot CEPs…with 2000-pound bombs."[65] One of the most significant supply chokepoints in North Vietnam remained: the Thanh Hoa Bridge. Attacked on 27 April with electro-optical guided bombs (the weather did not permit the launching of the flight's LGBs), it received several hits from the 2,000-pound Mk 84s, but it remained open for business. On 13 May, with better visibility, 16 F–4s armed with 15 LGB Mk–84s, 9 guided 3,000 pound Mk–118s, and 48 unguided Mk–82 bombs succeeded in dropping a span and wrecking the remainder.[66] The Seventh Air Force calculated that to inflict the same damage with unguided bombs, no fewer than 2,400 would

Figure 12. The Thanh Hoa Bridge, the "Dragon's Jaw", goes down May 1972. The Seventh Air Force study concluded: "It would have taken 2,400 unguided bombs" to accomplish the same degree of damage.

have been required.[67] The Vietnam War still holds the record for the greatest number of PGMs used in combat — more than 24,000.

The lesson learned from the employment of LGBs in Vietnam was that good results can be achieved in good weather when a man designated the target during the weapon's flight. What was also learned was that in the case of a hardened target, whether it be the overengineered Thanh Hoa bridge or an underground bunker, the munition must have an adequately sized warhead and an accurate guidance system.

What is also interesting to note is that the art of aircraft gunnery and bomb delivery had improved dramatically since World War II and Korea, as the following table shows. By the time Vietnam heated up, a great deal of training went into teaching munitions delivery as it related to the science of windage, aim points, and offsets. Not only did we have munitions-related education, we also developed aircraft systems with constantly computed impact points to assist greatly in that endeavor. Nothing, though, came close to the precision that was available with a munition that had terminal guidance.

Hit Probability of 90% against a 60' x 100' Target with the Use of 2,000-Pound Unguided Bombs at Medium Altitude[20]

War	Number of Aircraft	Number of Bombs	CEP (in feet)
WW II	9,070	3,024	3,300
Korea	1,100	550	1,000
Vietnam	176	44	400

Guided and Unguided Bomb Usage in Vietnam[69]

	Guided	Unguided
	Bomb Type	
Total Number of Bombs	26,690	3,476,000*
Total % of Bombs	0.2	99.8
CEP	23	447
Percent of Bomb Direct Hits	55	5.5
Percent of Strike Aircraft in Fleet Guided-Munition Capable	1	

With the end of Vietnam involvement, research on guided munitions did not follow historical precedent and come to an abrupt halt. Unlike the period after World War II and Korea when guided weapons lessons were ignored, some post-Vietnam lessons were taken to heart. The development of even more sophisticated and capable LGBs assumed a high priority in the Air Force. Based on the impressive accuracy of the Paveway I, the next major advance came

Figure 13. F–111 with GBU–28 "Bunker Buster."

with the GBU–16 Paveway II of the early 1970s with its folding wings and improved guidance.[70] Unfortunately, congressional cutbacks in the FY1974 budget request forced the Air Force to divert funds from other programs to keep the Paveway program on track. The low-priority GBU–24 Paveway III was initiated in 1976.[71] In the late 1970s, various high-tech weaponry programs, including PGMs, received Undersecretary of Defense for Research and Engineering William J. Perry's support, and the Air Force made a corporate commitment to the Paveway program. As such, the GBU–24's future was assured, and it first saw combat in the 1991 Gulf War, making an impressive showing. Another PGM made an impressive showing during Desert Storm, not only because of its ballistic effect, which was massive, but because of the speed with which it was fielded.

Shortly after Iraq invaded Kuwait in August 1990, U.S. intelligence sources began to verify and list the relatively large number of key hardened targets. To attempt their elimination but still ensure the survivability of the delivery aircraft, the size of the weapon would have to be severely limited owing to internal carriage requirements. But two days after Desert Storm began, with air superiority in hand, we could fly where and when we wanted, with what we wanted. As such, the internal carriage requirement disappeared for the demanding bunker-busting jobs on the air tasking order, and the externally carried bomb could be heavier and longer to achieve the required terminal effects.[72] In discussion with various Air Force Research Laboratory (AFRL) Munitions Directorate personnel, the name that emerges as the individual who

was the inspiration for what came to be the GBU–28 Bunker-Buster is Al Weimorts.

Weimorts was long aware of the informal requirement for such a weapon. With the informal requirement in hand, he made an observation and proposed a question: "We've got a BLU–82 that is a big bomb with a 500-foot CEP. What would happen if we put a precision guidance package on it?" That was the inspiration for the bunker buster, though Weimorts goes to great lengths to minimize his role and point out that everyone contributed significantly to the team effort.[73] As Weimorts himself described it, "people were lining up to help in any way they could."[74] Fabrication began on 1 February 1991, and the bomb was dropped over Iraq on 27 February — a record twenty-seven days.

The 4,700-pound GBU–28 Bunker Buster was developed specifically to attack buried Iraqi bunkers. The warhead was machined from surplus 8-inch Army artillery tubes that were recommended and procured by a technical lead from the Lockheed Corporation.[75] Guidance was provided by a Paveway III LGB kit with minor software changes.[76] A team of government and industry people came together striving for the resolution of a difficult technical challenge, and they did so in record time.[77] To be able to pull it off, they had to use as much off-the-shelf hardware as they could, and the only item that was really new was the warhead itself, the BLU–113 penetrator, with the rest of the GBU–28 system already existing in other forms (guidance software, and nose and tail kits). The BLU–113 warhead, based on the BLU–109 warhead, was made of high-grade forged steel with a narrow body that testing proved enhanced the warhead's penetration. The BLU–113 target penetration capability quadrupled that of the BLU–109.[78] Two were ultimately employed in Iraq.[79]

Superficially, some may infer from the GBU–28 story that the lesson to be learned is that science and technology programs take too long to transition to acquisition programs and that the R&D investment is unnecessary because a product like GBU–28 can go from concept to employment in twenty-seven days. The truth of the matter is that the long-term historical investment in basic research, modeling technology, and hardware development (warhead, explosives, fuzing, and guidance) under the LGB program allowed us to respond with a state-of-the-art point-design very quickly.[80] These research programs reach back to the early 1970s when AFATL (the forerunner of the current AFRL/Munitions Directorate) developed in-house simulation capability for all LGB variants, and these initiatives continued through the 1970s when Munitions Directorate analysis led to the eventual successful testing and acquisition of the GBU–10 (MK–84), GBU–12 (MK–82), and the GBU–16. This was also true for the Low Level LGB simulation developed to support System Project Office development, acquisition, and testing, which was ultimately used to analyze performance and modify the guidance law and generate delivery data used in all Desert Storm LGBs. The bottom line was that the technical data and knowledgeable personnel were in place to support the development of the required guidance and control aspects of the GBU–28.[81] Though the GBU–28 was, for the most part, off-the-shelf, the on-the-shelf items had a thirty-year heritage, as did

all the LGBs used in the 1991 Gulf War.

Just as there were guided munitions lessons learned from Vietnam, significant lessons came from the 1991 Gulf War. You cannot fly electro-optically guided weapons in sandstorms, thick oil smoke, or heavy weather. These drawbacks were resolved in another PGM program that possessed a long heritage as well: the Global Positioning System (GPS)-guided munition.

With the initiation of the GPS program in 1973, AFATL began work to develop an affordable GPS-aided inertial guidance and navigation munition package. This R&D ultimately resulted in an affordable technique for GPS-equipped weapons by the early 1990s, but it was not early enough for Desert Storm. The dramatic use of "smart" bombs during the 1991 Gulf War, which many observers thought revolutionary, was nothing more than a refinement of the guided weapons employed twenty years previously in Vietnam. Much of the GPS technological heritage can be traced directly to revolutionary breakthroughs in basic research.

In 1964, the Air Force Office of Scientific Research (AFOSR) initiated a program to develop and test a coded transmission technique that would provide precise ranging and target data. This Code Division Multiple Access System would allow all satellites in a constellation to broadcast on the same frequency without interfering with one another. In 1967, AFOSR grant research resulted in a conceptualized design for a low-cost GPS. These initiatives were joined with related U.S. Navy programs and came to fruition at a DoD conference in 1973. Today, GPS is a constellation of twenty-four satellites that provides instantaneous location and navigation information to all U.S. military services. The system has become second nature as a tool in reconnaissance, search and rescue, and especially in targeting. It is also significant to note that the work of two earlier Air Force–funded basic researchers was essential in making the GPS a viable system. In 1958, Rudolf Kalman perfected a mathematical algorithm filtering system that permits the routines that determine GPS positioning to update system parameters and to zero in on the correct satellite range and azimuth. Charles Townes's DoD-funded work on the maser in the late 1950s and early 1960s led to superior atomic clocks integral to the operation of every GPS satellite: the more precise the GPS signal, the more precise a GPS munition will be. Basic and developmental research resulted in the GPS constellation's being used as a foundation for a quantum increase in bombing accuracy—especially in the case of the Joint Direct Attack Munition (JDAM.)

Given the results and limitations of the LGB arsenal as demonstrated during the 1991 Gulf War, the Pentagon directed that a new weapon be designed to overcome LGB shortcomings: a day/night, all-weather, GPS-based smart weapon. In addition, it had to utilize existing air-delivered munitions' warheads (the MK and BLU series) as an economy measure.[82] To satisfy this requirement, an AFRL Munitions Tiger Team developed an Inertial Navigation System (INS)/GPS-guided all-weather weapon. This initiative leveraged earlier work under the Operational Concept Demonstration High Gear program, which eventually produced the JDAM.[83] In its ultimate form, the JDAM

is a $20,000 bolt-on kit, typically attached to a 2,000-pound free-fall bomb that, using position updates from orbiting GPS satellites, can maneuver to its target by the adjustment of its tail fins.[84] Whereas laser-guided weapons require that a person be part of the decision-making process (an individual to aim the laser), the JDAM is preprogrammed to hit a particular GPS coordinate. The Tiger Team decision to add an INS capability is an additional guarantee of success. If the GPS signal is not available, the JDAM's INS can be programmed from the launching platform.[85] The JDAM does have some drawbacks: its typical 13-yard CEP is not as accurate as an LGB, and its effectiveness against moving targets is problematic.

Thanks to years of basic, applied, and advanced research, the JDAM ultimately proved its exceptional worth in the weather over Yugoslavia/Kosovo during 1999's Operation Allied Force, and subsequently in Afghanistan and Operation Iraqi Freedom.

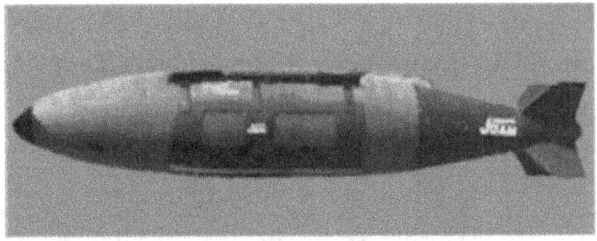

Figure 14. The JDAM has been the result of a long heritage in basic, applied, and advanced research.

JDAM Research Timeline[86]

Basic Research

- Adaptive Neural Net Control (theory), mid 1990s
- Modern Control Theory AP (theory, algorithms), late 1980s
- Optimal Trajectory Shaping Guidance Laws (theory, algorithms), 1980s
- CFD (algorithms), 1980s

Applied Research

- Triservice IMU (HG–1700), 1989–1992
- Tactical GPS Guidance (Kalman filter integrated with INS/GPS), 1990–1993
- Adv Adaptive Autopilot (neural network autopilot), 2001
- Weapon Integration and Design Tech (compact (compact wing kits), 1997–1999

Advanced Development

- Operational Concept Demo (INS/GPS), 1990–1991
- AGTFT (antijam), 1995–1998
- JDAM ER Demo (wing kit), 2000

In addition, scientific research has further enhanced PGM delivery with advances in the INS, a critical element of the JDAM.[87] Historically, all PGMs have had an INS. To be more precise, most PGMs have an Inertial Measurement Unit (IMU) that consists of three gyroscopes and three accelerometers which output changes in angle rate and acceleration. The output of the IMU is fed to a guidance processor that makes use of a Kalman filter. It is at this point, in the guidance processor, that the navigation data is computed: the latitude, longitude, and altitude. The IMU can be loosely, tightly, or ultratightly coupled to an update system like the GPS. A baro-altimeter can also be added. In fact, it can be said that the heart of the navigation system of most PGMs is the inertial system. A few PGMs like the early GBU–15s used only rate gyroscopes and had no inertial system. PGM guidance with only an IMU was first demonstrated at Eglin in a program called Inertially Guided Technology Demonstration in the early 1980s. This program demonstrated that an inertial system on a PGM could be aligned to a higher quality INS, launched from an aircraft, and autonomously guided to a designated target.

In the early 1990s, subsequent to the Inertially Guided Technology Demonstration program, Air Force Materiel Command Commander General Ronald W. Yates initiated the Operational Concept Demonstration High Gear in which Eglin demonstrated integrated GPS/inertial guidance on a GBU–15. This Team Eglin effort resulted in the integration of an IMU and GPS receiver as well as a navigation processor. Following the concept demonstration, Team Eglin produced a differential GPS version in a program called Exploitation of Differential-GPS for Guidance Enhancement in which, using a wide-area GPS network, they demonstrated very accurate GPS guidance against surveyed targets.

The current state-of-the-art IMU, integrated with a GPS receiver, is being flown on JDAM. It is projected that future IMUs will replace ring laser gyroscopes with MicroElectroMechanicalSystems (MEMS) devices. These MEMS devices (both gyroscopes and accelerometers) are currently in development. Tests are being conducted at Eglin demonstrating a prototype MEMS IMU in a JDAM. Thus it is that current systems such as JDAM, Joint Standoff Weapon, Wind-Corrected Munition Dispenser, Joint Air-to-Surface Standoff Missile, and cruise missiles all employ inertial systems, so their value cannot be overlooked, and the DoD research community will continue to enhance this valuable system.

The result of this R&D, to include basic research, has been the rise to preeminence of PGMs as the air-to-ground attack weapon of choice. In Vietnam, only 0.2 percent of the bombs used in were precision-guided; in Desert Storm, the figure stood at about 9 percent. During 1995 in Deliberate Force over Bosnia, of the 1,026 munitions dropped, 708 were PGMs, accounting for 69 percent of the total. In Operation Allied Force in 1999, fully 80 percent of the bombs dropped were PGMs.[88] With regard to operation Allied Force, Lieutenant General Dan Leaf provides a personal perspective of PGM employment in combat and some lessons learned.

> During operation Allied Force, I was fortunate to command the 31st Fighter Wing at Aviano, and during that my time there, we really expanded the flexibility of precision munition employment of Laser Guided Bombs. In particular, using them against mobile and fielded forces. My personal example is representative of such: In April 1999, I was flying on a FAC [forward air control] mission. My wingman and I spent an extended period of time looking for targets and we finally found a Serb convoy that was racing towards a town, and in our sense of the battle, was racing towards the town seeking sanctuary from us. We tried to stop the convoy with a couple rockets that my wingman fired, but this only prompted the convoy to increase their speed. My aircraft was loaded with two 500 pound laser-guided bombs. But I did not have time to set up a laser guidance pass, and in the interest of time, I dropped them both manually. One missed, because I did not have enough lead; and on the second, the reattack, I pulled enough lead and hit the lead vehicles of the convoy. Now, I lucked out and accomplished a pretty precise bit of bombing with a convoy moving somewhere between 30 and 60 miles an hour, and hitting them on the run. But it was really quite simple—you just have to pull enough lead to hit a moving target and you have to aim in the right place. But the fact that I elected to employ them unguided, because there wasn't time to set up the guidance, led me to have to make two passes, and led me to take significantly more risk. So when it comes to lessons learned, our PGM enhancements have to do with not only tightening the circle, but it is also very much an issue of tightening the kill chain by improving the employability of the PGM—making them more responsive, making them less operator intensive, while still allowing for execution of the operator's intent. Increased target processing speed begets improved accuracy begets greater situational awareness which results in more bombs on target, and a much safer mission.[89]

All of this also relates to time, which is a critical aspect of the PGM system. Time is not just speed. In PGM employment, it is the correct time, defined as when you will achieve the most desired effects, being able to respond when a target of opportunity presents itself,

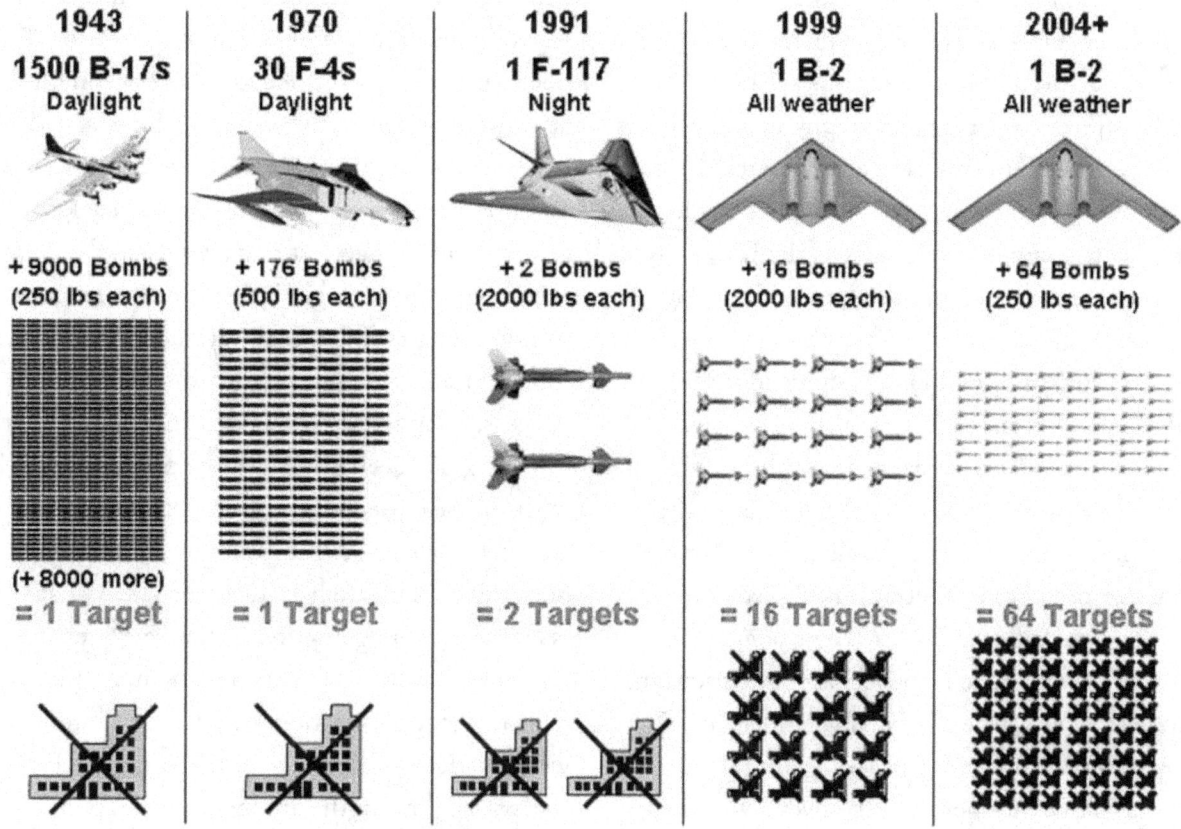

1943	1970	1991	1999	2004+
1500 B-17s Daylight	**30 F-4s** Daylight	**1 F-117** Night	**1 B-2** All weather	**1 B-2** All weather
+ 9000 Bombs (250 lbs each)	+ 176 Bombs (500 lbs each)	+ 2 Bombs (2000 lbs each)	+ 16 Bombs (2000 lbs each)	+ 64 Bombs (250 lbs each)
(+ 8000 more) = 1 Target	= 1 Target	= 2 Targets	= 16 Targets	= 64 Targets

Graphic and data (1943–1999) courtesy of Dr. David Mets, School of Advanced Airpower Studies, Maxwell AFB, AL; 2004 data courtesy of AFRL/MN, Eglin AFB, Florida.

assisting in a ground attack or other joint operation, or when you will negate or at least minimize any unintended consequences or collateral damage. We tend to think of PGM speed as being the be-all and end-all, but while faster is indeed better, time is a very complex element of precision. It is the correct time, whether it be the immediate moment or having a sensor system provide information to allow use of the weapon to its best advantage. And the way we will arrive at the use of future PGMs at just the *right* time is through research.

Research is the foundation for future weapons systems, and it is the driver behind enhancements of existing weapons. For example, an AFOSR basic research program grant contributed significantly to PGM accuracy when new mathematical architectures for control algorithms were used in field upgrades of the JDAM. The result was a significant increase in precision, as JDAM was declared the most effective munition recently used in Afghanistan and Iraq.[90]

Other areas where improvement can radically enhance PGM performance include the development of the Small-Diameter Bomb (SDB). In 1995 the commander of Air Combat Command asked the AFRL Munitions Directorate if a way existed to increase the bomb payload on our fighter aircraft so that more munitions could be carried to target with a commensurate reduction in the number of sorties required.[91] The munitions scientists' and engineers' starting philosophy was to use a small warhead, in this case, a 250-pound bomb with only 42 pounds of explosive, that would provide the same effectiveness as the JDAM, but with enhanced guidance. If one could improve the precision of a small bomb, plus improve its penetration, one would have a force multiplier. A fighter weapons platform with numerous small bombs could surpass the capability of two larger, in this case JDAM munitions. The resulting SDB owes its capability to a tritonic shaped nose to aid penetration coupled with guidance laws that zero out "angle of attack on impact" to optimize its penetration. In addition, these smaller warheads are independently targetable, allowing the attacking aircraft to ripple a large number of SDBs over a wide target area, with each bomb seeking its own (World Geodetic System 1984) loaded target location. Through range extension technology, which includes its compressed design, the SDB flight envelope has increased at least three times over JDAM distances. The result is that, from a single launch point, a B–2 can launch up to sixty-four independently targeted SDBs, thus becoming a force multiplier that can reduce sortie rates and enhance platform survivability.[92] Future versions of the SDB may have extended-range capability with pop-out wings and the ability to loiter or autonomously seek out targets.[93] PGMs that seek out their targets are now on the drawing board and offer unique capabilities.

With regard to autonomous loitering Automatic Target Recognition (ATR) platforms now in development, fielded results must be driven by the desired concept of operations and capabilities put forth in Air Force operational doctrine. ATR, in particular, is a very complex issue with numerous variables, and it is a system that may be easily countered. Clearly, autonomous loitering PGM systems face key challenges that have yet to be resolved, the critical one being target identification. Targeting can be thought of as a fifth variable in the four-element, precision-guided-munition-time process. Throughout the history of PGMs, targeting has been critical to the process, but with the proposed introduction of autonomous weapons, proper identification of the target becomes an absolutely critical burden: What is it that you are going to hit? Currently we rely on before-the-fact intelligence, not on real-time target identification, except when humans visually identify targeting opportunities. Thus, the critical link in the autonomous weapons platform will be identification to prevent unintended error. To emphasize just how imperfect this part of the process is, one need only be reminded of the JDAM attack on the Chinese embassy in Belgrade on 7 May 1999 during Operation Allied Force. Throughout Allied Force, the JDAM was employed only from B–2s, and it was very effective; for example, it hit precise intersections at targeted airfields. But the B–2/JDAM combination was also used in the strike on what we did not at the time know was the Chinese embassy. In this case, the JDAM hit exactly where it was supposed to hit. It went exactly where it was told to go, but it was told to go to the wrong place. The lesson here is that precision weaponry is much more than precision delivery: it is precision identification. Getting a PGM precisely to the wrong location can undo what a thousand PGMs accomplish that get precisely to the right location.

One of the state-of-the-art ATR systems now in development must address this challenging targeting issue, and a significant amount of past and ongoing research will determine its final capabilities. Like their predecessors, ATR systems have a rich heritage that derives from significant basic research discoveries and the enhancement of proven technology—primarily miniaturization, improved data processing speeds, and munitions enhancements.

Figure 15. *The autononous loitering Automatic Target Recognition (ATR) platforms now in development may share very little, from technical standpoint, with the 1918 Kettering Bug, but it is a direct decendent nonetheless.*

Autonomous PGM Research Timeline[94]

Basic Research
- LADAR ATR (algorithms), 2000–present
- Modern Control Theory AP (theory), early 1980s
- Multimode Warhead (phenomenology, theory), early 1980s
- Cooperative Attack (theory, concepts), 1999–present
- Agile Autonomous Control (theory, concepts), 2000–present

Applied Research
- Advanced Submunition Warhead Technology, 1995–1998
- LADAR Technology, 1994–2002
- Automatic Target Recognition Algorithms, 1994–2002
- Small Turbojet, early 1990s
- Multimode Warhead, mid-1980s–early 1990s
- Triservice IMU (HG–1700), 1989–1992

Advanced Research
- LOCASS (glider-armor), 1991–1994
- LORISK (SEAD/TBM), 1995–1997
- PLOCASS (powered), 1999–2002
- MALD engine, 1996–1998

The ATR PGM is only part of the future for PGM platforms. New and varied key technologies promise much more. Munitions now on the drawing board can be described as

"O" technologies: nano, info, bio, robo, micro, meso, and macro.[95] These technologies support the drive toward miniature systems with the potential for innovative payloads that enhance maximized load-out of delivery platforms and lead to low collateral effects.[96] For success in these effects-based targeting initiatives, R&D will focus on five areas: compact energy sources; mobility technology; miniature guidance technology; operations/communications technology; and manufacturing technology. With that as background, we can briefly explore the issue of future PGM development by keeping in mind two observations: we are now getting very target-specific, and a PGM doesn't always have to be something that blows up.[97] In fact, one could characterize this newapproach as a niche-based targeting system. These airborne-delivered PGMs would contain a sophisticated guidance system and a very tailored terminal effects package. For example, the weapon would arrive at its target area assisted by advances in aerodynamic shaping based on smart materials and structures and utilizing MEMS. Plasma or morphing technology would optimize the cruise conditions of an air vehicle based on flight regimes and atmospheric conditions. Once on the ground, the delivery vehicle itself would release a multitude of small PGMs that could themselves either fly or maneuver on land and arrive at preprogrammed destinations to perform reconnaissance or weapons functions based on immediate or future requirements. Further out on the technology horizon is an air vehicle that, once it lands, would not release seeker PGMs, but would itself morph into an entirely different vehicle utilizing many-structured or biomimetics-based materials to become a multifunctional PGM. As far as future terminal effects are concerned, once they arrive at their target location they could be electronically debilitating, structurally corrosive, an auditory or olfactory weapon, or reconnaissance-specific.[98]

The guidance and control of PGM weapons is the critical issue. Political, environmental, and moral considerations, not to mention critical warfighting requirements, place great demands on our scientific munitions-related community. To answer these demands, the basic research community is providing the required building blocks based on nanoelectromechanical and microelectromechanical systems, biomimetics, and intelligent systems research. For example, future nanosystems will provide accelerometers and gyroscopes based on micromachined inertial measurement devices. MEMS research will provide the required microswitches for superior low-cost millimeter-wave antenna elements. Biomimetics, by reverse engineering biological processing systems (using insect models) will develop new image-processing techniques. Intelligent systems research will support autonomous munition development

such as ATR, which allows a PGM to choose the correct target from a set of potential aim points. Today, much of this research has transitioned to the applied research arena.[99]

One area under development, which will be an outgrowth of guidance and control initiatives, concerns Micro Air-Delivered Munitions, which allows small munitions to adapt to their surroundings in a relatively crowded urban environment or be able to penetrate cave or underground complexes. These micro-platform systems could operate covertly to navigate, sense, map, reconnoiter, and attack targets from behind enemy lines. Specific missions would include the neutralization of critical underground nodes to compromise facilities containing hazardous chemical, biological, and nuclear weapon storage. They could also be used in an area dominance role to aid in searching a target area and identifying and defeating a specific target. In a bomb damage assessment role they would collect and transmit data for weapon detonation assessment in real time. Some of the many warfighter benefits that would accrue from the insertion and use of micro weapons and munitions include improved target intelligence, covert disablement of targets, remote mapping, long-residence time-on-target, controlled collateral damage, and a cost-effective weapon system.[100]

Not only warfighter benefits accrue from scientific advances in guidance and control. Dual-use opportunities are numerous: mapping, medical imaging, data compression, vehicle identification, collision avoidance, autonomous landing systems, commercial navigation, environmental monitoring, and intelligent highway systems, to name but a few.[101] The fact is that the U.S. scientific community does not own the monopoly on advanced technological solutions. Take, for example, the existing unique lattice, grid-fin steering assemblies employed on some current and future PGMs. These box kite–like fins enable a large amount of lifting surface to be packaged in a compact space, and they have an interesting background. They would not have been possible without the developmental grid fin-work accomplished by Russian and Soviet scientists over the past forty years, who ultimately shared them with their munition designer counterparts at the Eglin munitions lab.[102]

In looking back over the long, evolutionary history of guided-munition development, one is indeed struck by what has been a 100-year process — from Karl Jatho's pilotless biplane of 1903 to the autonomous PGM-equipped weapons in development today. While much has changed, much has not. For instance, PGMs will not replace the requirement for an area-bombing capability. Area bombing is not the indiscriminate evil some portray it to be. There will continue to be large target areas such as facilities, fielded forces, and other nonspecific targets that could be better exploited with area munitions. A related issue dealing with the moral aspect of area bombing is an even greater moral obligation and burden placed on U.S. forces with regard to PGM employment. Given the television showcased capabilities of PGMs in recent conflicts, the American public, and to an even greater degree, the rest of the world, may be utterly convinced that anything we bomb, we bomb on purpose, that there are no accidents of war, that if we hit a nonmilitary target, we were aiming for that target. As such,

this is one aspect where our PGM success has highlighted our failures. Our success is a two-edged sword, and we must be constantly aware of both edges because heightened success in the employment of PGMs will only increase the pressure to perform always to perfection. We will, though, look to the continued results of the scientific community to help us toward that end, always striving to better perfect this new "revolution in warfare" that they made possible, and it will be a continuing revolution.[103] What Lieutenant General Buster C. Glosson wrote in 1993 is still valid today:

> *Any way you cut it, we will need smart airplanes with smart weapons to meet the challenges of the future. However, during this period of frenetic change, we would do well to remember King Solomon's counsel that "wisdom is more important than the weapons of war." He is right. People are always more important. All the so-called smart weapons in the world could not distinguish their own tail fins from the Pentagon if it were not for the smart people who develop, build, maintain and program them [and] they are already developing the weapons we will need to win the next one.*[104]

Enlisting the Spectrum for Air Force Advantage:
Electro-Optical/Infrared (EO/IR)

Ruth P. Liebowitz

Edward A. Watson

Major General Stephen G. Wood, USAF

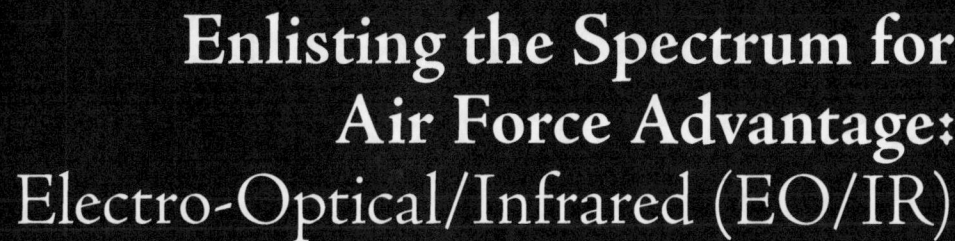

Abstract

The electro-optical/infrared (EO/IR) wavelength in the spectrum has provided a versatile sensing tool for military use since at least the 1950s. EO/IR imaging technologies have contributed significant capabilities to meet Air Force requirements for detecting, identifying, and tracking military targets. The EO/IR has proved to be a versatile sensing tool for many applications since the 1950s. Although each system and platform utilizing EO/IR sensors requires a different design, all share some basic technical parameters and components.

This paper describes the EO/IR technologies developed to achieve current Air Force capabilities. It focuses on infrared thermal imaging systems and places the development of these technologies in the context of changing political-military conditions. It also includes a discussion by an Air Force operator of his experience with an EO/IR application of growing importance: sensors on unmanned aerial vehicles.

INTRODUCTION

Seven years ago we could not begin to imagine the level of impact that placing an electro-optical/infrared (EO/IR) camera on a medium-endurance, remotely piloted vehicle would have on combat aviation. The Predator weapon system has heralded a revolution as significant as any since the Wright brothers' first flight. It is through imaging technologies that we have been able to expand our situational awareness and capture knowledge for real-time use as well as for postflight analysis. The unique characteristics of the EO/IR spectrum have provided us with very precise information upon which we have based a myriad of decisions. The unmanned aerial vehicle (UAV) known as Predator (see figure 1) provides a microcosm wherein we can see the synergy of imaging technology and aviation.

Figure 1. The Predator.

The current success of the imaging systems onboard the Predator is based on a long history of research and development. For the last fifty years, the Air Force has funded broad research on remote sensing in the visual wavelengths of the spectrum and beyond the visual, that is, the EO/IR range, in order to gain this kind of advantage.

EO/IR technologies emerging from this research have been developed into sensor systems. These systems provide the Air Force with a powerful tool for detecting, identifying, and tracking military targets. For many years, these sensor systems were a force multiplier for fighters and bombers before they assumed a dramatic new role on UAVs. This paper will give special attention to one subset of electro-optical systems, namely infrared cameras, because they have contributed a key capability for the military the ability to conduct nighttime operations and because their realization has involved major technical challenges.

THE HISTORICAL CONTEXT

The original impetus for research and development programs in the EO/IR area came from the experience of World War II. The leaders of the Army Air Forces were impressed with the importance of science in the victory of the Allies. They cited the key roles of radar and the atomic bomb, to name two of the most important products of science enlisted for the war effort. In planning for the future, Army Air Forces leaders organized a program of long-range research and development to continue to exploit the advantages of scientific knowledge to help meet military challenges of the future. Late in 1944 the Army Air Forces' General Henry H. (Hap) Arnold observed to his science advisor, Dr. Theodore von Kármán, As yet we have not overcome the problems of great distances, weather, and darkness.[1] For the new Air Force (as of 1947), these deficiencies translated into efforts toward achieving three ambitious goals: a greatly extended range for air operations, an all-weather flying force, and the ability to conduct round-the-clock operations. This last goal, nighttime capability, provided a clear rationale for investing in research in the EO/IR wavelengths of the spectrum. EO/IR systems, along with radar and lasers, became the three main prongs in remote sensing for military applications. Support for research in these areas was also informed by a more general goal, namely, to limit U.S. casualties in warfare. This axiom, as articulated by General Arnold late in 1944, stated: "It is a fundamental principle of American democracy that personnel casualties are distasteful. We will continue to fight mechanical rather than manpower wars."[2]

The priorities of the Cold War reinforced this initial impetus toward advanced technology for military capability. Ongoing concern about numerical superiority in the Soviet bloc sustained the strategy of reliance on the countervailing resources of science and technology. In the 1990s, the first post Cold War decade, this strategy continued, but it was now augmented with information superiority to provide an asymmetric advantage against threats posed by terrorism and regional instability. Infrared remote sensing offered promise for operations in total darkness, a capability much sought-after during the Vietnam War as a means to interdict enemy resupply efforts and supply routes. Infrared remote sensing also had a role in plans to hold off potential assaults from the Soviet bloc. During the first Gulf War it provided a tool for dealing with mobile Scud missile launchers. Its ability to identify hot spots in aircraft, tanks, motorized vehicles, and facilities (see Figure 2) is of continuing

importance for targeting, as is its ability to be used in battlefield conditions of smoke or dust (though not in rain or fog). Since infrared cameras are passive systems, they do not transmit signals that an enemy could utilize for targeting. Though infrared imaging systems lack the vast detection range and all-weather capability of radar, their strengths, when used for reconnaissance missions, are higher resolution and more naturally appearing images. Therefore, they have found application in the identification and verification of targets close-in.

Figure 2. *Thermal imagery of a truck and a van. Since a small amount of infrared light depends on the temperature of the object, brighter features represent hotter parts of the object.*

Air Force supported work in infrared technology has focused on developing capabilities in navigation, targeting, and reconnaissance as a force multiplier for airborne operations. Since the later 1940s, laboratories run by the service have conducted ongoing research programs to investigate materials and designs suitable for infrared imaging systems, to specify atmospheric conditions relevant to "seeing" at different wavelengths, and to solve technical challenges in the processing and display of infrared data.[3] Historically, airborne infrared programs have been centered at Wright-Patterson AFB's Avionics Laboratory, with significant work done at Hanscom AFB's Electromagnetic and Geophysics Divisions and at Griffiss AFB's Rome Air Development Center (RADC) (later named Rome Laboratory).[4] In the current, consolidated Air Force Research Laboratory, most of these programs now reside in its Sensors Directorate.

TECHNICAL CHALLENGES

To achieve the desired capabilities for the warfighter, a number of important technical challenges had to be met. The development since World War II of detector technology for infrared systems has been a long and difficult, though eventually highly successful, process. As with radars and lasers, the design of the instrumentation requires carefully considered tradeoffs to optimize the system for different operational configurations. Sophisticated image processing done by computers is sometimes desirable to enhance particular features of the image and, in some cases, necessary to compensate for limitations in the optical instrumentation. With time, efforts have focused on combining infrared, radar, and laser remote-sensing technologies in suites of sensors to augment their capabilities and allow greater operational flexibility. Studies are currently underway on integrating and fusing data from different remote-sensing instruments. Another major technical challenge has been the realization of techniques to improve the discernment of targets in the infrared against their natural backgrounds.

Much progress has been made in this area since the 1960s, and the results of this research have been packaged for ease of use by the operators. Some aspects of the infrared military research have given rise to significant spinoffs for scientific and commercial use.

DETECTOR TECHNOLOGY OF INFRARED CAMERAS

The development of detector technology for infrared imaging systems has emerged in roughly three stages, with full flowering not attained until the 1980s and 1990s. Before this, in the early 1950s, the Air Force had adapted a direct, heat-seeking infrared technology for use with air-to-air missiles on fighter aircraft. The early versions of the AIM9 Sidewinder, which was first fired successfully in September 1953, typically had only single detectors and did not form an image per se; rather, the receiver contained a spinning reticle that modulated the signal the detector saw. This modulation was used to guide the missile to its target. Later versions of the Sidewinder, however, began to use infrared imaging technologies to make the missiles more robust from longer ranges.[5]

It is the detector technology that has essentially defined the three generations of infrared cameras. The cameras share the characteristic of all electro-optical systems of being sampling imagers, in contrast to the continuous vision of the human eye or of typical film cameras. Electrical signals generated by the detector are relayed to some form of computer for processing and display. Sampling methods, however, have been quite varied. First-generation systems generally consisted of a single detector. Scan mirrors were used in the optical system to scan the detector over the required viewing angle. Usually one mirror would scan rapidly in the horizontal direction while a second would scan more slowly vertically. A complete scan of the viewing angle had to be made within the time allotted to a single frame, about one-thirtieth of a second, thus limiting the number of samples that the imager could make. This also made the imager less sensitive because the detector could not gather large amounts of light by staring for a very long time at a single point in the scene.

Second-generation infrared cameras made use of linear arrays of detectors, that is, a line of detectors as shown in Figure 3. Many points in the scene could be sampled simultaneously. Use of a linear array greatly reduced scanning requirements, since only the single, slower scanning mirror was required to scan the array in one direction. The sensitivity of the array was also better because each detector could stare at a point in the scene for a longer period. This infrared camera typically had two fields of view: the wide field of view was for target acquisition; the narrow field of view provided a magnified view of a smaller portion of the image, providing better resolution for target identification. Mounted on fighter aircraft with a forward view for navigation and targeting, their orientation gave them the nickname FLIRs (forward-looking infrared systems).

The Avionics Laboratory at Wright-Patterson was primarily responsible for developing this generation of infrared cameras as part of its broader work on technology for targeting and weapons guidance in air-to-ground scenarios.[6]

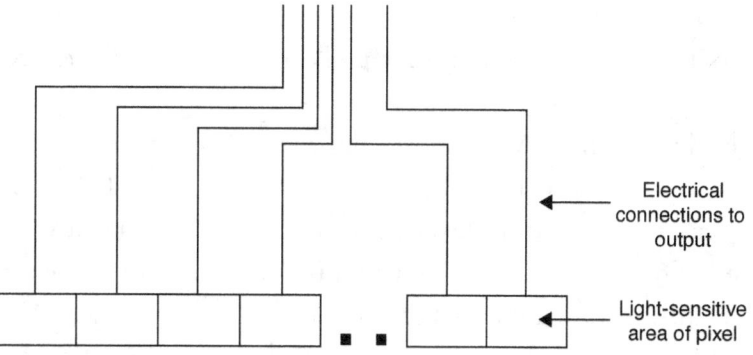

Electrical connections to output

Light-sensitive area of pixel

Figure 3. Schematic of a linear detector array. The detectors can be electrically connected to the output circuitry through their free sides.

FLIRs were an important component for the smart weapons that emerged in the decade after the Vietnam War. Seekers for an air-to-ground missile such as the AGM65 Maverick meant a television camera in its early versions, but later versions employed infrared cameras to allow performance at night and better operation through battlefield obscurants such as smoke and dust. FLIRs were frequently used in conjunction with laser designators for bombs equipped with laser guidance systems. One challenge accompanying this new technology was to get it integrated into fighter avionics and platforms. Sometimes the sensors were placed inside the airframe; more often they were suspended below it in pods. Another related challenge, especially in the years when FLIRs were first introduced, was to acquaint pilots with the new technology and train them so they were comfortable with using it. By the 1980s, the Air Force's main fighter aircraft like the F4E were carrying EO/IR systems such as Pave Tack along with their munitions. The newer Low Altitude Navigation and Targeting Infrared for Night, or LANTIRN as it is usually called, which was under development in the 1980s, was a two-pod system (see Figure 4). LANTIRN's AAQ13 navigation pod contained a wide-field-of-view FLIR and a terrain-following radar. Its AAQ14 targeting pod included a targeting FLIR, a laser designator/ranger, an automatic tracker, and an automatic handoff of the target to a Maverick missile.[7]

Third-generation infrared cameras are known

Figure 4. LANTIRN – An F–16C/D Fighting Falcon fighter plane with LANTIRN pods mounted externally beneath the body of the aircraft. On the left, the targeting pod; on the right, the navigation pod.

as staring-sensor systems. They made use of two-dimensional arrays of detectors (Figure 5) in which each pixel could now stare at a single point in the image for an entire frame time. This made the third-generation system more sensitive. In addition, the need for a mechanical scanning system was eliminated, making the optical system less complicated and less expensive to maintain. The intent of these new cameras was to achieve a parallel technology to home video-camera technology at infrared wavelengths. This approach had been tried earlier in the first two decades after World War

II, but it failed because of an inability to achieve sufficient contrast in infrared scenes. The new design, first proposed by Air Force scientists at the Electromagnetic Division at Hanscom and at RADC in 1973, led to a pioneering platinum silicide camera, the first staring camera that could be used effectively for infrared thermal imaging.[8] The camera held the promise of being able to detect very small variations in temperature and to yield images with better resolution. Key to its success was the use of the new combined material, platinum silicide (PtSi), for the detector substrate. It offered characteristics that could help offset the difficulties with nonuniformities in the arrays and other sampling issues inherent in staring systems. By the 1990s, infrared cameras were able to detect extremely small variations in temperature and provide high-resolution images.[9]

Improvements to the staring camera have

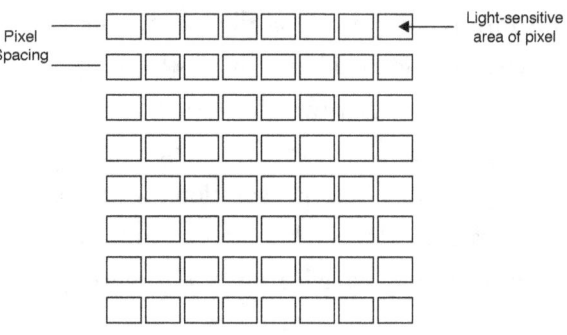

Figure 5. Schematic of the detector array used in the electro-optical imaging systems. Each detector is made of a light sensitive material in the wavelength region of interest. The size of each detector (small is better) and the number of detectors in the array (larger is better) affects the quality of the image produced by the array. Arrays for visible light tend to have much smaller pixel sizes and much larger numbers than arrays for infrared light.

continued, leading to further improvements in detector materials and in techniques for dealing with nonuniformities in the array and sampling issues. Additional research on other materials such as indium antimonide and mercury cadmium telluride has produced infrared cameras with increased sensitivity at many different wavelengths of the infrared. A technique called microscanning has been developed to overcome the effects of sampling. In the microscan process, several images are collected, but each is shifted a fraction of a detector element with respect to the other images. By appropriately recombining these images, a greatly enhanced image can be produced.[10] The nonuniformity of staring arrays can be dealt with to some extent by post-detection processing, which produces an enhanced final image for display. To compensate for the variations in each detector element, periodic calibrations are performed and are applied after the image is detected to remove the effects of nonuniformity. Additional post-detection processing can include techniques to remove blurring and remove other effects associated with sampling by the detector array. Some of these processing techniques are quite computationally intensive and can be applied only after the imagery has been gathered. Other techniques are simpler and can be applied as the imagery is gathered and displayed to the operator in real time. The increasing capability for processing images has also enlarged the options for their display to the operator. For reconnaissance applications, the images have tended to be transmitted as stills. When the processing before display keeps pace with the sampling rate of the camera, however, a stream of images with seeming continuity can be created. This mode of transmission has led to creative new operational uses of the technology.

The extent of technical achievement in infrared technology can be appreciated by comparing infrared cameras with the technology for electro-optical systems that use visible light. The latter include such devices as low-light cameras and night vision goggles, electronic cameras, and television. They use low-level visible, or near visible, light, such as moonlight or starlight as an illumination source, and then amplify it. In contrast to these visible systems, which work off reflected light, infrared cameras operate by receiving thermal light emitted by an object on the basis of its temperature (see Figure 2). The development of infrared cameras has been more technically challenging in a number of ways. Because the common glass used for fabricating the optics for visible systems does not transmit in much of the infrared region, designers of infrared systems have had to use other materials. These can be more difficult to work with, possibly being water-soluble or toxic. Similar issues exist with the detection component. Detector arrays for visible light are typically based on silicon, the material from which computer chips and most other electronics are fabricated. After many years of huge commercial investment, silicon is a very well-understood material. Military applications for low-light cameras have leveraged off this investment. In contrast, the materials for infrared detectors are typically made from material alloys such as mercury-cadmium-telluride, platinum silicide, and indium antimonide. The military has devoted long-term resources to research alloys, but the utilization of alloys has not reached the same level as the utilization of silicon. Lastly, unlike low-light cameras, the detectors for infrared cameras and their surrounding area have to be cooled to very low temperatures to reduce background emissions. Only with recent advances in cryogenic technology has this become less of a practical issue.

By the 1990s, operational versions of PtSi cameras were being installed on Air Force airborne platforms, including the B52 bomber fleet.[11] They also became elements of suites of sensors for new platforms being developed. An array of next-generation sensor systems, including electro-optical systems primarily in the infrared spectrum, was reported in the planning for the Joint Strike Fighter.[12] The two endurance UAVs, the Predator and the Global Hawk, which the Air Force has been managing since the mid-1990s, utilize the latest versions of infrared cameras. The Predator, a medium-altitude endurance UAV, flies under 25,000 feet. It carries electro-optical systems configured for this altitude, tightly packaged into a small space. For the first version of the Predator to go operational, the suite of sensors included a synthetic aperture radar and a combined EO/IR camera. The video footage from the sensors is sent down live, in real time.[13] The Global Hawk, still in development and acquisition, is a longer-standoff platform that flies at high altitudes (greater than 60,000 feet). This platform can carry several different types of sensors including synthetic aperture radar and visible electro-optical and infrared sensors. The infrared sensor is based on an indium antimonidestaring array sensitive in the midwave infrared band. The visible sensor

is a Kodak detector array based on silicon.

FUTURE DIRECTIONS

The capability of electro-optical systems continues to advance. Research in active electro-optical systems is receiving considerable attention. These can complement the use of infrared cameras under obscured visual conditions, such as clouds, dust, and smoke. In electro-optical systems, a laser is used as an illumination source. It is analogous to conventional radar in which radio frequency energy is directed toward an object and the backscattered energy is collected. A primary advantage is its ability to range gate. A laser pulse is directed toward the object of interest, but the camera in the receiver is kept off while the pulse travels to the object and the scattered light returns. Just before the pulse returns to the receiver's optical system, the camera is turned on. It therefore captures light reflected only from the object of interest and not light scattered from any intervening obscurant. While these obscurants do reduce the amount of light that is returned to the receiver, they do

Figure 6. Example of laser-illuminated imagery. Note that because the laser sends out the light in pulses, only the region about the object is illuminated. This allows the camera to ignore scattered returns from intervening obscurants such as smoke and dust.

not affect the image resolution greatly (Figure 6). Other features of active electro-optical systems include the ability to measure many different features of an object, including its vibration, color, and surface characteristics.[14]

TARGET/BACKGROUND CONTRASTS

In fielding these generations of electro-optical and infrared systems, one major technical challenge has been to understand the atmospheric channel through which images of targets present themselves to the optical system. The passage of light can be affected by attenuation because molecules and particulates constituting the atmosphere absorb or scatter the propagating light. The attenuation at some wavelengths is so great that it prevents their use in electro-optical systems. Similarly, atmospheric optical turbulence the same process that causes stars to twinkle can also blur images produced by imaging systems. To improve the operational use of these systems, Air Force researchers at the

Air Force's Geophysics Division at Hanscom developed a set of atmospheric transmission models based on atmospheric properties and meteorological variables. The models, which have become standard Department of Defense (DoD) and industry codes under the names LOWTRAN and MODTRAN, give transmittance/radiance predictions for electro-optical and infrared systems.

In this area also, infrared imaging systems pose particular problems. The infrared target/background contrasts for an infrared image can be quite variable. The image is determined by a set of sometimes complex relationships involving temperature differences between the target itself and objects surrounding it, and the background is determined by atmospheric and meteorological conditions. Unexpected reversals of anticipated target/background contrasts can sometimes occur (see Figure 8). Early in the Vietnam War, the Air Force took measurements of remote sensing in the infrared in Thailand

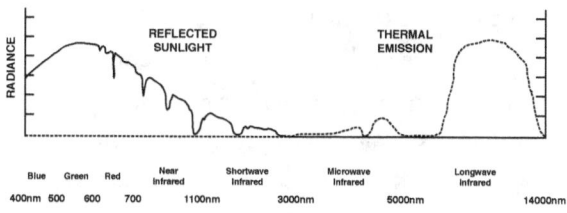

Figure 7. The light reflected and emitted from a "typical" object is shown here. The wavelength scale at the bottom has been altered to move easily show the visible portion of the spectrum. Note that most of the emitted light is in the long and mid wave infrared, while most of the light in the visible portion of the spectrum is due to reflected sunlight. The transmission characteristics of the atmosphere are impressed on the curves. The lack of light near 3,000nm, 5,000-8,000nm, and at various portions in the near and short wave infrared is due to atmospheric absorption in these regions.

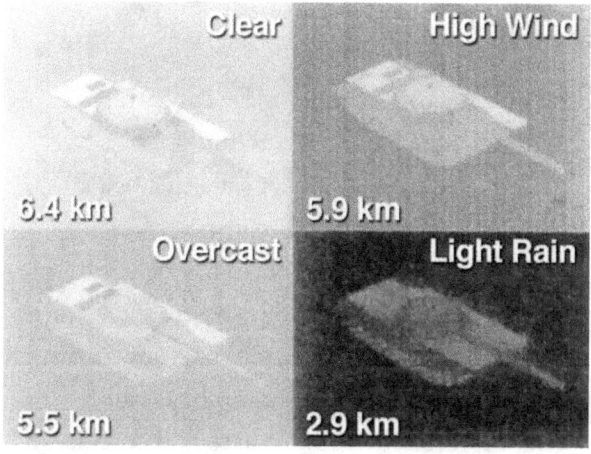

Figure 8. A T–62 tank against a soil background presents a very different image under clear, overcast, windy and rainy conditions. Variations in the optical and meteorological environment for a mission result in different target acquisition distances for a specific electro-optical system.

to assess the expected performance of early sensors in the tropical jungles of Southeast Asia. Similarly, when military attention returned to Central Europe in the 1970s, an extensive set of measurements was made there. In the 1980s the atmospheric transmission models were combined with weather and infrared target/background data to create a set of electro-optical tactical decision aids. These tools enabled pilots to compare maximum acquisition range for a variety of smart weapons under given atmospheric/meteorological conditions so that they could select the best munitions package for a mission.[15] Another tool both for operations and for training simulated scene generators was developed to assist Air Force operators in visualizing their targets in the infrared. A current version of this tool, the Infrared Target-Scene Simulation Software, was recently adapted for use in urban target areas of interest in Iraq.[16] The extensive research conducted in the EO/IR area has produced effective technologies that maximize system performance.

SPINOFFS

The detector technology for the infrared staring cameras has some nonmilitary applications, particularly in scientific fields that require low-cost and high-precision measurement systems. One of the most exciting uses is in infrared astronomy where the infrared staring cameras provide a tool to study regions of star formation.[17] The technology is also being explored for possible uses in medicine. In industrial applications, a variety of applications employ thermal imagers, including process control, monitoring of high temperature components, energy management, checking for

leaks in systems, and identifying weaknesses in structures. Thermal imagers are also finding their way into the automotive industry. A few models of cars now have a thermal imager as an option which can be purchased to achieve improved road awareness at night. Because of cost considerations, these imagers usually employ simpler detectors that do not have to be cooled.[18] The atmospheric research that supports the software for maximizing the performance of FLIRs has been widely utilized for civilian and commercial purposes. In particular, DoD's atmospheric transmission code, LOWTRAN, has become ubiquitous.

CAPABILITIES FOR AIR FORCE OPERATIONS FROM VIETNAM TO KOSOVO[19]

Early electro-optical and infrared imaging systems saw some utilization in the Vietnam War and then played a much more significant and expanded role in the first Gulf War. Beginning in Bosnia and continuing in Kosovo, Afghanistan, and Iraq, these systems have become central to the new capabilities in reconnaissance and battle management.

In the Vietnam War, early versions of the AIM–9 Sidewinder missile were used extensively by fighter aircraft, and their later versions have been part of the inventory in succeeding campaigns (see Figure 9).[20] The AC–130 gunships in charge of search and rescue operations in Vietnam carried what was very sophisticated equipment for the time: low-light-level television cameras, and early infrared sensors. Infrared detection was considered as the most promising technology to permit combat aircrew recoveries at night. A limiting factor, however, was the technology's inability to distinguish between the body heat of a downed aircrew and that of an enemy soldier.[21] RADC installed airborne infrared systems to provide pilots with instantaneous data during reconnaissance flights. It also trained C47 gunship aircrews in the use of FLIRs. RADC's base security programs in the Vietnam era utilized the infrared as one of their technologies.[22]

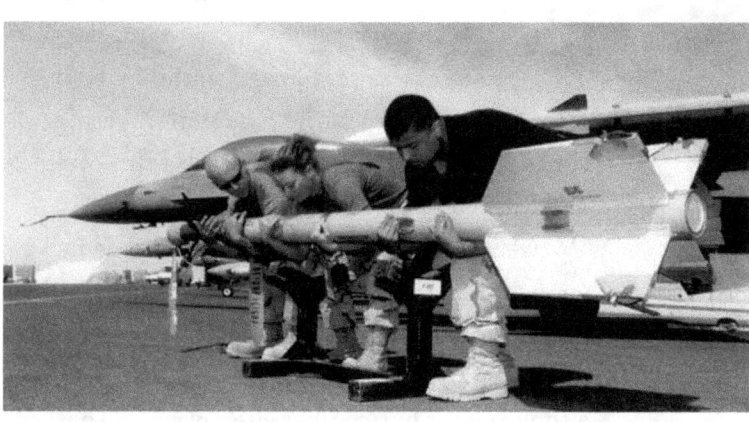

Figure 9. From left, Airman 1st Class Bradley Smith, Staff Sergeant Jessica German and Airman 1st Class Gerardo Gonzalez hoist an AIM–9 Sidewinder missile before mounting it to an F–16 Fighting Falcon aircraft. The airmen are assigned to the 379th Expeditionary Aircraft Maintenance Squadron at a forward-deployed location in Southwest Asia supporting Operation Enduring Freedom.
(Photo by Master Sergeant Terry L. Blevins)

By the time of the first Gulf War, a group of mature, infrared imaging systems had entered the Air Force inventory. There were the PAVE systems and advanced infrared versions of the AGM65 Maverick (Figure 10). Moreover, the new LANTIRN pods came into the operational inventory in the later 1980s and were carried on the F15E Strike Eagle and the F16C/D Fighting Falcon. They allowed the fighters to

fly at low altitudes, at night, and under the weather to attack ground targets with a variety of precision-guided and unguided weapons. Like the Joint Surveillance Target Attack Radar System, or Joint STARS as it is usually called, LANTIRN was conceived as a counter to Soviet and Warsaw Pact capabilities and was then pressed into service in the Gulf War before it was fully ready for operational use. During the Gulf War, infrared imaging systems were used to good effect under conditions of adverse visibility.

The capabilities of the LANTIRN targeting system resulted in the Tactical Air Command giving LANTIRN-equipped F15s the mission of finding and destroying mobile Scud missile launchers.[23]

As of the mid 1990s, the latest electro-optical systems were to be seen on Air Force platforms beyond manned fighters, bombers, and gunships, marking the start of a new era. Equipped with video and infrared cameras, the Predator UAV began to assist with monitoring operations in Bosnia late in 1995 while still an Advanced Concept Technology Demonstration. The Global Hawk UAV (Figure 11) did not make its debut until 2001 in Operation Enduring Freedom (OEF) in Afghanistan.[24] Although the Predator was originally a Navy program, its sensor capability has been expanded since 1996 under Air Force management, and it has been put to dramatic new uses in situational awareness and battle management. These operational breakthroughs depended on the availability of advanced satellite communications technology.

In its monitoring operations in Bosnia and

Figure 10. AGM–65 Maverick Launched from an F–16 fighter. Courtesy of Raytheon Company.

THE PREDATOR IN OPERATION ENDURING FREEDOM: AN OPERATOR'S COMMENTS[26]

The integration of modern EO/IR sensors on the Predator enabled us to follow every significant movement of our enemies in Afghanistan and Iraq. With the use of traditional intelligence collection, it could take hours or even days to process an image, assess that image, and then task a strike mission against the potential target. In times past, our adversaries observed this cycle and learned to move its equipment to survive. We were always one step behind and only lucky enough to destroy a portion of the targets we planned each day. This chess game of move and countermove provided our enemies with a tool to limit our combat effectiveness. The introduction of streaming video has changed the rules of the game and forced our enemy into a constant state of defense and fear. The use of live EO/IR feeds has provided a decided improvement and subsequent advantage. Our enemy has lost the sanctuary of time between when we observe his actions, assess the military viability of the target, and execute a strike.

The original Predator aircraft, RQ1, used off-the-shelf cameras and commercial data links to send the picture instantaneously (we refer to this technology as streaming video), first to the pilot and sensor operator and then to anyone given the link receiver. Air Force Chief of Staff General John Jumper, then Commander, United States Air Forces in Europe, was handed this

then in the air campaign over Kosovo in the spring of 1999, the Predator was increasingly deployed in tandem with other Air Force platforms. In Bosnia, the Joint STARS E8 aircraft cued the Predator to execute closer searches in difficult terrain. The use of a long-flying, unmanned vehicle, one that was inexpensive enough to be considered as an expendable, allowed much greater freedom in reconnaissance and surveillance. Later in Kosovo, both the Predator and the high-altitude U2 reconnaissance plane were electronically linked to Joint STARS. All were coordinated through command and control centers in order to conduct more effective search-and-destroy missions.[25] The Air Force was to make an even more creative and far-reaching use of the Predator two years later in Afghanistan.

Figure 11. *Global Hawk unmanned aerial vehicle.*

capability to augment his operations, including the tactical search for targets in Bosnia. The full-motion, real-time, in-your-face video converted him to a UAV believer. General Jumper understood the real value of this capability and envisioned it could do much more than just watch the enemy. He saw its ability to not only locate potential targets, but to guide weapons to a precise location or provide other weapons systems with precision location data in a killerscout role, even laser-designating the target for weapons delivered by other platforms. As the conflict was winding down, he took the first steps in fulfilling this vision by having the Predator outfitted with a laser targeting system aligned with the cameras. The rest of the vision was fulfilled as General Jumper saw the fielding of a complete weapons system when the Hellfire missile was integrated into the Predator (Figure 12). The fulfillment of this vision was a natural evolution of technology and operations.

Predator aircraft are designated two ways: the RQ1 has a sensor ball equipped only with cameras; the MQ1 has a laser integrated with the cameras. We use the RQ1 with its 20-hour mission time to search for and track specific enemy targets. During the day, the primary cameras used are the electro-optical cameras with the 16- to 160-millimeter Zoom lens and the 955-millimeter fixed Spotter lens. This gives a National Imagery Interpretability Rating Scale (NIIRS) 7 rating at the 5 nautical mile slant range. Operationally this means I can track a specific a person or vehicle in moderate traffic beyond the range in which they can hear or see the Predator. Integrated in the sensor ball is the infrared camera capable of seeing in the 3 to 5 micron range. It has a six-stepped Zoom from 11 millimeters to 560 millimeters. This gives a NIIRS 4 rating at the 5 nautical

Figure 12. *Demonstration of a Predator-launched Hellfire missile blowing up a tank.*

mile slant range. During the day this camera gives us the ability to find hot targets. It is the camera that watches through the night.

The live video of the war produced by these cameras brought extraordinary situational awareness and was quite addicting. While I was the night director for the Combined Air Operations Center at Prince Sultan AB during OEF, I had the video displayed on the wall-sized situation display we called the Big Board. The capability to capture and display this time-critical information was invaluable to the decision-making process and was often the focal point for all those not engaged in critical taskings on the operations center floor. This was truly the first reality show.

While all our systems and airmen performed admirably, we could not have done our job as well as we did without the Predator. The ability to watch the Toyota truck transporting Al Queada fighters as it drove through the city or watch in real time the rescue of the Christian Aid workers was indispensable to fighting the war. There was and continues to be more demand for Predator than we can possibly fulfill. We are striving to put the necessary resources in place to meet the need, but demand often outstrips availability. The Predator system has become a must-have capability. The demand for Predator support is pervasive. I was not the only one who absolutely relied on the video feed. For certain decisions, the National Command Authorities and often the President himself watched the video feed.

Throughout the course of OEF, success brought new concepts of operation, which brought even more success. Before the war we worked hard to train our aircrews to direct

fighter attacks on enemy positions. In one small city in Afghanistan, Predator crews located an Al Queada cell meeting in one section of a mud-brick complex. Even with precise knowledge of the target location, dropping a large conventional bomb could produce significant collateral damage and possibly kill a number of innocent civilians. To avoid this potential, an AC130 gunship was called in to take out the apartment. In the middle of the night, the Predator pilot [sic] walked the gunner's eyes from the town square, down narrow twisting alleyways, past numerous look-alike buildings, to the specific room, which was then precisely attacked. The surgical-strike capability of the gunship combined with the acute situational awareness and precision of the Predator had a tremendous synergistic effect. On another occasion, a truck with enemy fighters fled from a building that was under attack. They ditched their car and began walking across fields away from the truck, thinking they were safe. It wasn't their day. Predator was watching every move they made, and one by one, our forces tracked them down.

While we were regularly successful, It took time and skill to talk an attack aircraft onto a specific target at night. Some incredibly smart people asked the question, Why can't we send the picture directly to the cockpit of the striker? This was the genesis of the Rover modification. Predator was modified to turn its line-of-sight antennae toward a specific AC130 or a receiver-equipped ground party. The person needing the information received the signal and was rewarded with awesome situational awareness.

During Operation Anaconda it was not uncommon for Al Queada fighters to hide in ambush, waiting for our advancing ground forces. As our special operations forces moved in, it looked like someone had kicked an anthill. Our enemy crawled out of their caves and hid among the trees to try to cut our forces down. Predator was watching and tracking each enemy tree. As our forces advanced through the woods, Predator video was sent directly to the AC130 to highlight which tree belonged to us and which to the enemy. One by one we dismantled the ambush and took the stronghold.

These are but a few of many successes that sparked the idea for another improvement, the MQ1 equipped with a laser designator and Hellfire air-to-ground missiles. All of you have undoubtedly seen the precise attacks Predator made. Military equipment hidden near mosques was taken out without disrupting the meeting going on inside. A specific car parked among other cars, one particular element in a group meeting at a camp, and one antenna on top of a building in Iraq were each precisely struck with no collateral damage. This is unprecedented accuracy never before seen in the history of modern combat. Predator has no unique technology, but the synergy of fielded technology has made it the poster child of revolutionary weapon systems.

Improvements are still in the works for the MQ1. We have purchased a new sensor ball with expanded capabilities that will generate even better employment tactics. The infrared camera is significantly better than what the RQ1 had onboard. Not only has the image fidelity improved, full zoom instead of step focal lengths will be featured. The new sensor ball adds low-light television sensitive to the near-infrared spectrum, nominally 0.7 to 0.9 microns. Predator will see the markers used in

the night vision range. Image processing has been added to give point or area track capability as has a coast function that will keep the camera moving to aid in reacquiring a target that has become obscured. The video display will also fuse all the cameras onto one picture. In the automatic mode, it will select exactly what percent of electro-optical, low-light television, and infrared will be presented. These changes will keep the system locked on the target longer, especially as conditions deteriorate.

The next-generation Predator, MQ9, will fly even higher and faster. The RQ1 typically flies at medium altitudes (~20,000 feet) at 80 knots indicated. The MQ9 will fly up to 50,000 feet in altitude at 275 indicated and will carry full-size weapons. The debate is ongoing over what camera system will be fielded. Initially, only still images are being targeted for collection. Based on operational experience with the MQ1, high-quality, affordable, motion video would be the operator's desire. I would challenge academia and industry to work together and deliver the technology that will enable our continued battlefield dominance through superior situational awareness and precision weapons delivery capability.

CONCLUSION

Since the 1950s, EO/IR technologies have evolved enormously. Infrared cameras, in particular, have achieved a key capability for the Air Force the ability to conduct operations in total darkness and this had a significant impact during the first Gulf War. Used in conjunction with radar and lasers, they provided an important tool for reconnaissance, navigation, and targeting. Upgraded FLIRs have been mounted on each new generation of the Air Force's fighter aircraft. To arrive at this accomplishment, extensive research in difficult materials was undertaken, and challenging design problems were surmounted. The increasing availability of computer software for sophisticated image processing has also been an enabler for infrared cameras. Similarly, supporting research into the performance of FLIRs under different climatic and meteorological conditions kept pace with U.S. engagements in different parts of the globe.

Thus, by the 1990s, EO/IR technologies had been greatly advanced and constituted a significant force multiplier to the Air Force's fighter, bomber, and gunship fleets. Although this development represented many steps forward and a few major breakthroughs, such as the creation of the PtSi staring-sensor camera, there was not a great element of surprise in this area until the mid 1990s when mature EO/IR technologies were incorporated into the Air Force's new UAVs. As this combination was being used in Predator operations in Bosnia and Kosovo, the realization of its potential for expanded situational awareness emerged. The potential, of course, depended on capabilities offered by the latest satellite communications technology. From this beginning came the glimpse of exciting new scenarios for missions in which targets could be very narrowly defined, and the whole sequence of the action observed and assessed close-in, without risks to personnel. This gave impetus to further technical development of the Predator   improving the EO/IR sensors, adding a laser for target guidance, and exploring the option of a missile to fire from the platform. At the beginning of the new century, the Air Force was beginning to realize some of these unanticipated possibilities in OEF, with an expectation of more to come.

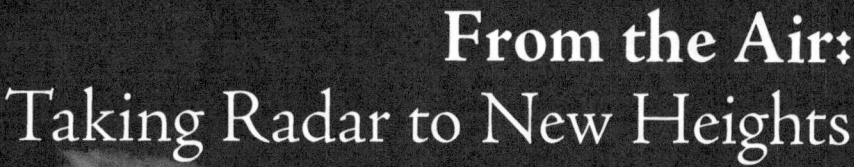

From the Air:
Taking Radar to New Heights

Tom W. Thompson

Ruth P. Liebowitz

Jon S. Jones

Robert J. Mailloux

Lieutenant General Bruce K. Brown, USAF, Ret.

Major General Joseph P. Stein, USAF

Abstract

The advantage of the vertical, that is, the capability to see farther and better than the adversary, has long been a cardinal principle of warfare. This paper explores how radar and sensor technologies developed in Air Force laboratories have permitted application of the principle in what is perhaps its most effective use to date: airborne remote sensing for purposes of command and control.

Beginning in 1954, the Air Force organized airborne radar early-warning units with EC–121 aircraft. The Airborne Warning and Control System, the Joint Surveillance Target Attack Radar System, and fighter-based airborne remote sensing systems followed. Key to our success were Air Force enabling technologies, particularly low side-lobe antennas, signal processing, airborne moving target indicator and ground moving target indicator systems, forward-looking solid-state radars, and sensor fusion. Numerous examples, most recently, Operations Enduring Freedom and Iraqi Freedom, attest to the operational utility of airborne remote sensing.

INTRODUCTION

Niccolò Machiavelli (1469–1527) thought a general should possess a perfect knowledge of the localities where he is carrying on war.[1] Today, United States Air Force airborne remote sensing[2] stands close to making that ideal a reality. Airborne remote sensing provides the means not only to seize and hold space over the field of battle but, even more significantly, to understand it in ways that permit exploitation in a manner Machiavelli and the generations of military theorists who followed him almost certainly never imagined.

HISTORY

The history of Air Force airborne remote sensing stretches over half a century, beginning when the Air Force became a separate armed service with its own laboratory and research and development (R&D) structure. Several technologies attracted Air Force interest during these early years, but radio detection and ranging (radar) stood out. A closely guarded secret during the Second World War, radar became a vital component in national defense strategy during the Cold War, and Air Force laboratories spared no effort in exploring ways to improve and apply the technology more widely and effectively. Research gained momentum in the wake of recent developments in military aviation, like jets, supersonic flight, and ballistic missiles, all of which required better and sometimes new means for surveillance, tracking, and target engagement. Air Force laboratories turned to klystrons,[3] traveling wave tubes, and more powerful magnetrons to boost radar performance. Other efforts focused on improving ways to process

radar signals, a major goal being to eliminate clutter, or unwanted radar returns. Lighter, more reliable components and moving-target indicators (MTIs) that eliminated blind spots and canceled electronic interference emerged as research priorities. Infrared and electro-optical sensors attracted research interest as well for their potential to buttress radar.

Symbolic of early Air Force airborne remote sensing was the EC–121 Warning Star (Figure 1) originally developed by Lockheed Corporation for the Navy. The Air Force, which designated the system RC–121, procured fifty-six of the piston-driven planes and used them to warn of approaching unidentified aircraft and ships.

Figure 1. EC–121 Warning Star aircraft.

Warning Star carried an S-band APS20 surveillance radar and an X-band APS4 height-finding radar.[4] By 1957, the Air Force had several squadrons of the aircraft in service.[5] Reliability problems plagued the EC–121's radar and electronics. However, improvements like data-link communications and increased radar range[6] improved reliability but ultimately proved inadequate. Reality was that more formidable targets and missions called for better technology. Moreover, the problem of ground clutter persisted, particularly over land, where it obscured targets and offered potential adversaries an opportunity to avoid detection.[7] In 1967, the Pentagon approved a modernization plan that called for over-the-horizon backscatter radar, a new interceptor, and an Airborne Warning and Control System (AWACS). AWACS, known as the E 3 Sentry, represented a significant advance over its EC–121 Warning Star forebear. An outgrowth of the Overland Radar Technology program, which developed pulse-Doppler techniques for rejecting radar clutter, its AN/APY1 radar was superior to the EC–121's APS20. The AN/APY1 was a look-down radar whose antenna operated from the top of a specially modified Boeing 707 aircraft

Figure 2. Airborne Warning and Control System (AWACS).

in an elliptically shaped rotodome 30 feet in diameter and 6 feet thick. With AWACS, the Air Force acquired the extended radar coverage and increased capability it had wanted to detect and track low-flying targets.[8] By 1977, Electronic Systems Center began transitioning the first of the aircraft to Tactical Air Command (TAC), and the AWACS achieved initial operating capability in April 1978. All told, thirty-four E–3A Sentries entered Air Force service.[9]

Besides new technology, AWACS introduced new operational concepts for airborne platforms.

AWACS, though a greatly improved airborne platform for radar-guided air-to-air intercepts, was more than that; it was also a command and control system: the AWACS would facilitate command and control of deployed tactical forces. Such tactical air operations as rendezvous, aerial refueling, air traffic control, interception, search and rescue, reconnaissance, counter air, battle damage assessment, direct air support and aerial resupply could be monitored and controlled by AWACS.[10]

New technology made such command and control possible, principally consisting of longer-range radar, better radios, digital communications, and computer displays. Enhanced command and control, including the means to monitor enemy air activity beyond the forward edge of the battle area, became possible.[11]

Meanwhile, more powerful computers and more precise navigational guidance had begun to presage the next round of airborne remote sensing advances. During the Vietnam War, Rome Laboratory[12] scientists and engineers devised ways to improve the targeting of enemy communications. Although the war ended before the technology could be placed on the battlefield, it was, perhaps, the most significant Air Force technology development of the period. The Advanced Location Strike System (ALSS) represented the state of the art, zeroing in on enemy electronic emissions for guided air-to-ground weapons. Designed to locate and strike enemy surface-to-air missile (SAM) fire-control radar and supporting communications, ALSS used ground control stations, beacons, and airborne relay platforms to direct airborne munitions. Employing coherent correlation and Kalman filtering techniques in a coherent-emitter location testbed, researchers tested ALSS at the White Sands Missile Range, New Mexico, and in Europe. The technology would carry over into other programs, eventually finding its way into Air Force and Army weapon systems of the 1980s and 1990s.

By the 1980s, such technology was underpinning a doctrine known as Assault Breaker which emphasized precision conventional munitions delivered by tactical aircraft and radars against second-echelon Warsaw Pact forces. The idea was to stabilize an initial Soviet Union assault against Western Europe and then bring it to heel by destroying its capability to replenish and regroup from reserve or second-echelon forces. Optimists saw the new doctrine as offering a possible means for ending conflict by using conventional weapons and tactics. Remote airborne sensing systems figured prominently in the strategy, especially the part that emphasized targeting enemy command and control and other surface targets from afar.

Pave Mover, an Air Force program that sought radar capable of picking out targets on the ground and directing munitions against them, and the Standoff Target Acquisition System, a U.S. Army effort to develop a helicopter-mounted MTI radar that detected enemy armor beyond the forward area of battle, illustrated the technological and doctrinal change (Figure 3). Working with the Defense Advanced Research Projects Agency (DARPA) and the Electronic Systems Division, Rome Laboratory used several technologies, most notably airborne phased-array antennas, synthetic aperture radar, high-speed analog/

digital converters, beam-steering computers, distance-measuring equipment, parallel signal processors, and sophisticated algorithms, to develop Pave Mover radar. The system utilized two communications links. One provided a surveillance picture to a ground station while the other sent target coordinates for munitions. The high point of the program occurred in 1982 when, during tests in New Mexico, Pave Mover radar 75 miles away guided a fighter aircraft as it dropped dummy weapons on a moving tank column the pilot never even saw.[13]

Both Pave Mover and the Standoff Target Acquisition System transitioned to the Joint Surveillance Target and Attack Radar System (Joint STARS) program. The Air Force had responsibility for the airborne part of the system, and the Army, the ground portion. In 1984 the Air Force selected the Boeing 707 airframe for Joint STARS. Like AWACS, Joint STARS was to operate from a modified Boeing 707.[14] Their similarities tended to end there. Although Joint STARS was also a long-range command and control system, its command control extended to mobile surface targets.

Figure 3. Standoff Target Acquisition System (SOTAS).

74

Housed in a canoe-shaped radome under its fuselage was a side-looking phased-array antenna capable of detecting, tracking, and targeting enemy ground forces, whether in motion or stationary (Figure 4). The radar possessed wide-area surveillance, MTI, and synthetic aperture capabilities. The result was an airborne command and control system capable of using aircraft to target surface targets up to 150 miles away.[15] Flight-testing of Joint STARS began on 1 April 1988;[16] it achieved initial operational capability in December 1997.[17] Eventually, the Air Force plans to operate a fleet of Joint STARS, with fleet size estimates ranging from twelve to as many as nineteen aircraft.[18]

Nor were airborne command and control platforms the only aircraft benefiting from ongoing advances in airborne remote-sensing technology. Air Force fighter and bomber aircraft benefited too. Lighter, more compact avionics permitted the use of more sensors per platform, which in turn increased operational capabilities.[19] Forward-looking infrared extended distances aircraft could search, and lasers allowed for more precise target-ranging and acquisition. By the 1990s, Air Force fighter and bomber aircraft carried sensors that, when combined with systems like AWACS and Joint STARS, greatly enhanced the effectiveness of the United States military (Figure 5).

The Affordable Moving Surface Target Engagement (AMSTE) program, which applies Ground Moving-Target Indicator (GMTI) radar along with high-speed processors, low-cost standoff precision weapons, and communications networks to the task of destroying mobile ground targets, seeks to improve on Joint STARS.[20] Unlike Joint

Figure 4. *Boeing 707 retrofitted with Joint STARS under the fuselage.*

Figure 5. *F–15 (above) and F–16 (below) carring sensors that, combined with AWACS and Joint STARS, enhanced the effectiveness of the United States military.*

STARS, which identifies targets and tasks other aircraft to track and destroy them, AMSTE monitors targets until their destruction. The heart of AMSTE is a computer processor that fuses data from two separate airborne ground-surveillance radars and simultaneously transmits the information to missile-firing aircraft.[21] The Air Force is currently evaluating AMSTE technology for possible use in Joint STARS, Global Hawk, the F/A–22 Advanced Tactical Fighter, and the F–35 Joint Strike Fighter.[22] Interest in the technology stems from what such airborne remote sensing potentially offers: the capability to find, strike, and destroy targets sooner, even if they turn, brake, or accelerate.[23]

Because the success of AMSTE depends on fusing data from different sensor and information systems, tools to accomplish this have received research emphasis. Such tools include automatic tracking that converts GMTI detection data into tracks; motion-pattern analysis that analyzes GMTI tracks for significance; behavioral pattern analysis that considers enemy intent; and resource and scheduling that commits multiple platforms against targets. Algorithms, particularly those that distinguish moving ground targets among objects on the battlefield, form a conspicuous part of all research. The objective is a familiar one: engagement of time-critical targets in near real time.[24] In parallel with these developments, the Aeronautical Systems Center performed upgrades and enhancements to the U–2's radar, giving it a higher resolution capability and an MTI mode. In 1996, development of an unmanned aerial vehicle (UAV) that carried a low-cost radar with both synthetic aperture and MTI modes began.

Impressive as each of the aforementioned airborne remote sensing systems is, their effectiveness depends on how well they work together. The reason is simple but critical: no one sensor suffices to meet the challenges of the modern battlefield. Rather, the situation demands many sensors operating as a single system. The Air Force describes such sensing as multispectral, a data-fusing technique that relies on many types of sensors, not just on radar alone. Multispectral experiments that tested concepts and technologies for fusing data from different types of sensors became common in Air Force laboratories during the 1980s and 1990s.[25] This fusion requirement created its own set of challenges, not the least of which was how to correlate data generated by different sensors operating miles apart among a variety of platforms.[26] In addition, this process had to occur in near real time for munitions to be targeted effectively against moving targets. The enabling technology in this instance was the electronic computer, which had grown increasingly powerful as it shrank in size. Both AWACS and Joint STARS relied on computers to correlate and track the data their sensors provided. During the 1990s, Off-Board Augmented Theater Surveillance research explored ways to integrate intelligence, surveillance, and reconnaissance (ISR) among Joint STARS, AWACS, and Rivet Joint aircraft to assist in target identification. Of particular interest was how such technology might be used to distinguish between friend and foe and to identify neutrals during the fog of battle. Work went forward under the aegis of Network Centric Collaborative Targeting, which planned to demonstrate fusion technology connecting the three aircraft with the off-board surveillance information in a series of future exercises.[27] Ultimately, what the Air Force wanted was automatic target recognition.[28]

TECHNOLOGY

That Air Force laboratories developed better radar and sensor technologies and then fashioned them into operational airborne remote sensing systems, while impressive, forms just part of the Air Force airborne remote sensing R&D story. The other, equally impressive part is how they allowed sensors to operate as a single entity. What characterizes modern airborne remote sensing is its dependence on a range of technologies, integrated and cooperating, to detect and eventually engage the target. This is no small accomplishment, given the challenge of integrating voluminous amounts of information from multiple sources at various locations. For decades, it was the long pole, as it were, in the airborne remote sensing R&D tent, and solving it took consistent commitment over many years. Sustaining that commitment was the understanding that future airborne remote sensing capabilities depended on obtaining solutions.

Although no single technology brought forth this fusion capability, solid-state electronics deserves much of the credit, principally because it made the essential computer and signal processing possible. Revolution may be too strong a descriptive, but the term comes close to acknowledging the significance of what was happening as solid-state integrated circuits became smaller, more efficient, and, most importantly, more economical to produce. The 1970s ushered in minicomputers, called microprocessors. Compared to the Electronic Numerical Integrator and Computer (Eniac), the first electronic digital computer, the typical microprocessor was 30,000 times cheaper, 300,000 times smaller, consumed 56,000 times less power, and performed 200 times more calculations.[29] The Air Force quickly grasped the significance of this and began developing integrated avionics systems that could be embedded in aircraft.

Programs like Digital Avionics Information System, Pave Pillar, and Pave Pace used solid-state components to integrate, improve, and reduce the size and weight of avionic architectures, including sensors. Solid-state components in radars made them more reliable, improved their capability to track targets in clutter, and gave them better countermeasures. Forward looking airborne radar, incorporating solid-state technology, performed more functions. It was not long before forward-looking radar possessed modes for navigation and weather, terrain following/terrain avoidance, precision target mapping (synthetic aperture mapping), air-to-ground weapon delivery, and countermeasures. This combining of multiple modes in radar with rapid electronic beam scanning became the basis for the modern solid-state, multifunction radar.[30]

Trends toward greater computerization and better signal processing carried over into antenna technology. Typifying these efforts was the Avionics Laboratory's[31] Molecular Electronics for Radar Applications (MERA), a solid-state module and antenna array. Initially, MERA's objective was to advance microwave integrated circuits, but it expanded to components and antenna arrays.[32] The MERA module and array had multiple radar modes for ground mapping, terrain following/avoidance, and air-to-ground ranging. The array had 604 elements, produced 32 dB gain on transmit and 30 on receive, transmitted 352 watts peak at 9 GHz, and had a system noise figure of 12.5 dB. Another Avionics Laboratory effort, the radome antenna and radio-frequency circuitry program, sought simple, lightweight, economical ferrite phase-shifter scanning arrays. An electronically agile radar followed. It employed mass-produced phase-shifter/radiator elements costing approximately $300 each, considerably less expensive than the technology developed under MERA.[33] The electronically agile radar influenced design of the B–1B avionics.[34]

Despite the achievements of these early passive arrays, they suffered reliability problems because of their continued reliance on tubes and high-power voltages. More solid-state technology became the remedy. The reason was obvious: solid-state array modules not only required less power, they delivered higher mean times between failure. Solid-state antenna arrays tolerated hundreds of module failures before needing repair, a fact that gave them a crucial edge over their tube-based cousins. The Avionics Laboratory's reliable advanced solid-state radar, developed during the 1970s, boasted an array of 1,648 elements, each transmitting 1.4 watts peak power at a 5 percent duty cycle. The reliable advanced solid-state radar achieved a mean time between failures of 27,000 hours based on one transmit or one receive failure.[35]

The solid-state, phased-array radar program took advances to new levels before ending in 1988. It used microwave integrated circuit technology to demonstrate even more convincingly the increased capability and reliability that solid-state antenna arrays provided. If there was a negative, it was cost, which remained relatively high compared to more traditional technologies. A breakthrough in affordability arrived with the availability of monolithic microwave integrated circuit technology, which DARPA began sponsoring as a means for improving advanced solid-state technology. Monolithic microwave integrated circuits and solid-state phased-array radar modules found their way into the ultrareliable radar, which would later support Advanced Tactical Fighter and F–22 radars.

The technology that transformed fighter and bomber aircraft produced equally significant changes in Air Force early-warning and command-and-control platforms. With AWACS, they manifested themselves in a pulse-Doppler airborne, moving-target indicator (ATMI) antenna (Figure 6). Since AWACS flew at approximately 30,000 feet, it needed an antenna that surveyed 360 degrees of azimuth while looking downward at airborne

Figure 6. A pulse-Doppler airborne moving target indicator (ATMI) antenna.

targets as low as 500 feet above the ground. Separating radar returns of airborne targets from radar clutter originating from the ground posed a daunting challenge. Very low antenna side lobes, along with much improved signal processing, provided the solution. Basically, MTIs eliminated ground clutter immediately beneath the target while the very low, side-lobe antenna suppressed clutter originating elsewhere. Doppler radar-filtering helped too, enhancing target signals. Actual target detection occurred in a pulsed-Doppler mode.[36] To determine elevation, the antenna employed a planar S-band array of horizontal rows of edge-slotted wave guides. The main beam measured approximately 1 degree wide in azimuth and 5 degrees wide in elevation.[37] Equally impressive was what was achieved in size and weight. The entire antenna operated from a 30-foot-wide rotating radome.

The basic antenna design, a slot array, was not revolutionary, but its precision signal control was. Precision stemmed from row-arrays that kept antenna side-lobe radiation low. The method, which owed much to Air Force Cambridge Research Center studies of computer codes for antenna interelement coupling,[38] proved essential in the design of the array.[39] Precision extended to measuring equipment too, making it possible to find and evaluate previously undetectable signal errors. Finally, computer-controlled milling machines cut the waveguide slots with unprecedented precision. It was the most precise, lowest, side-lobe antenna array yet built.[40]

Upgrades throughout the 1980s and 1990s ensured the continued viability of AWACS. New and more powerful computers and more sensitive radar components boosted the system's performance. A fiberglass radome reduced

weight; new software permitted tracking ships; and additional radios and consoles made communications more reliable.[41] Later enhancements allowed detection of targets traveling at less than 85 knots and its operation with the Joint Tactical Information Distribution System. Designed to eliminate security problems associated with voice communications, this joint distribution system provided computerized communications links between AWACS and the forces over which it exercised command and control.[42]

The Joint STARS radar antenna was even more advanced, its phased-array design and signal processing permitting detection from the air of slow-moving ground targets in clutter.[43] Designed to operate in a displaced phase center mode that compensated for the motion of the aircraft, the antenna processed radar returns while canceling clutter and tracking ground targets. The antenna, a planar array of 456 X-band slotted waveguides, was 24 feet wide and 2 feet high. Each waveguide slot array provided a low side-lobe fixed beam for elevation. Signal processing combined signals from all 456 waveguides to form either a narrow beam pattern or three wider beams. Phase-shifters kept antenna side lobes low in all modes and distributed power across the antenna aperture with unprecedented precision. Unlike the AWACS array, which had very low side lobes in a plane of the array that did not scan, the Joint STARS array formed low side lobes in all planes, including the plane of the electronic scan. Phase-shifters also distributed power in a way that set the amplitude for the narrow beam as well as the three wider displaced beams.

Enhancements to Joint STARS throughout

the 1990s boosted overall capabilities. New satellite communications links improved capabilities to transmit Joint STARS data through the Air Force chain of command. A computer replacement program reduced the number of onboard main computers from five to two and increased processing power, speed, and reliability with a fiber-optic local area network technology.[44] New algorithms played an essential part in upgrades too, principally as a means to give Joint STARS the ability to distinguish more rapidly targets from vast amounts of data. Research included algorithms for rotating-antenna identification, MTI data synthesis, off-board data cueing and correlation, high-range resolution, convoy detection, motion pattern analysis, and modeling. A moving-target exploitation effort got underway to develop automatic tracking of targets, which included combining high-range resolution radar with MTI to produce a single dimensional radar cross section of a moving vehicle. That these advanced algorithms supported open computer architecture designs made future changes easier to make. Subsequent development of Moving Target Information Exploitation made much of this moving-target information available to ground stations. The addition of World Wide Web capability allowed Internet technology to be used for exploiting data. Early in 2001, Joint STARS, U–2, and Global Hawk aircraft shared GMTI data in an experiment.[45]

Such developments, particularly the increasing role of algorithms, underscored the critical link between remote airborne sensing and computer software. Software told computers what to do, and, as computers improved, demands for software capable of exploiting them to the full increased. For airborne remote sensing, computer software became essential to efforts seeking to computerize more and more radar functions, a trend that continues. Driving this trend is the opportunity to shift some of the burden of further improvement from radar and other sensor systems to computer processing, which, potentially, improves performance and cuts costs. But software had to be made affordable, a tall order given the relatively labor-intensive characteristics of software writing and engineering. By the mid 1970s, the Department of Defense was spending three times as much for computer software as it was for computer hardware.[46] Costs had to come down if the Air Force was to realize the full potential of the computer. Not surprisingly, the Air Force worked the software cost issue intensively throughout the 1980s and 1990s, developing systems that automated formerly manual processes associated with software production. Computer programming languages improved, and processes associated with software development got automated.[47] Software costs fell, and computer processing became even more embedded in Air Force systems.

OPERATIONS

Research in Air Force laboratories, while it focused on long-term objectives, always had a more immediate concern: the warfighter. This became apparent during war or national crisis when priorities inevitably shifted to supporting current operations. Airborne remote sensing was no exception, a fact exemplified in the case of the EC–121 Warning Star. Besides serving as an early-warning system of bomber attack against the United States, it became, during the Vietnam War, an airborne, tactical aircraft radar-control facility. Radar weapons controllers on Warning Star coordinated with forward air controllers to deliver air strikes on enemy targets, effected midair refueling with tankers, assisted in search and rescue, and helped recover aircraft to home bases.[48]

In 1966, Warning Star began supporting Igloo White, a program designed to interrupt the enemy's use of the Ho Chi Minh Trail as a route for resupplying its forces in South Vietnam. By the close of 1967, parts of the Igloo White system were in place. Essentially it sowed the trail with innumerable electronic sensors. These sensors, relying on seismic, acoustical, and chemical data to determine the presence of the enemy, tracked attempted infiltrations into South Vietnam. EC–121 aircraft, flying above, relayed the received information to a computer-monitoring station where planners interpreted it.[49] The computer station, located at Nakhon Phanom AB, Thailand, processed the information and gave it to strike aircraft for immediate attack.[50]

Like its EC–121 predecessor, AWACS came to play an important air defense role. North American Air Defense Command assumed control of AWACS in 1979, which involved keeping a significant portion of the aircraft on alert. The AWACS had other roles too, in particular with the North Atlantic Treaty Organization (NATO) forces in Europe. From 1978 until 1989, AWACS, flying from Keflavik, Iceland, provided long-range surveillance of and airborne intercept control against Soviet bombers. In Europe itself, AWACS supported United States Air Forces in Europe (USAFE) in its mission to deter the Soviet Union from

launching a surprise attack on Western Europe.[51] All told, eighteen AWACS deployed to Europe. One contingent, based at Geilenkirchen, Germany, was operated by NATO crews.[52]

AWACS mirrored Warning Star in that operators, ultimately, determined the success or failure of the system. Lieutenant General Bruce Brown, for example, remembers how AWACS skeptics in TAC became believers when, during an exercise, they witnessed what the system working with F–101, F–102, and F–106 fighter aircraft and AWACS radar weapons controllers could do in a tactical scenario: Not a bad night; 199 kills out of 200. The same guys who said it was too expensive, was too easy to jam, and couldn't survive (among their many criticisms) were now trying to convince me what a marvelous machine we had on our hands.[53] Similarly, General William Creech has recalled how operators, during exercises, played an essential role in AWACS's acceptance: We had a lengthy demonstration in Europe of the AWACS for the NATO brass while I was USAFE Director of Operations, Intelligence an initiative that culminated in the NATO AWACS program. The System was so impressive that it sold itself.[54] As for those, who, in the early 1970s, had considered AWACS too big, too slow, too expensive, and too vulnerable, he had but one reply: dumb.[55]

Military contingencies also revealed AWACS' worth. Urgent Fury, the 1983 United States operation against Granada, provides an early case in point when AWACS were pressed into service, helping direct and control air operations. Then, in 1989, AWACS supported Operation Just Cause in Panama. Less than two years later, during the Persian Gulf War, AWACS again saw action. During Desert Shield, the buildup of United States and Coalition forces in the Persian Gulf, AWACS deployed to Saudi Arabia to guard the skies against possible Iraqi attack. Later, during Desert Storm, the ground invasion by United States and Coalition forces that expelled Iraqis from Kuwait, AWACS controlled the air-to-air war and coordinated the hunt for Iraqi Scud missiles.[56] AWACS saw action again during Operation Deny Flight, the operation designed to keep Serbian aircraft grounded during the Bosnian Crisis.[57] In 1999, during Operation Allied Force, AWACS directed and controlled NATO air strikes against Serbian forces in Kosovo. Fourteen AWACS aircraft flew 656 sorties in support of NATO operations.[58] AWACS also supported Operations Enduring Freedom and Iraqi Freedom.

Joint STARS received its baptism of fire even before it became fully operational. The occasion was the Persian Gulf War of 1990–91. During Desert Shield, General Norman Schwarzkopf, Commander in Chief of Central Command, ordered two Joint STARS aircraft to deploy to Saudi Arabia. The two aircraft took turns monitoring Iraqi armor and troop movements. During Desert Storm, Joint STARS flew over the battle area, providing information on Scud missile sites and giving intelligence about Iraqi troop movements. Joint STARS proved particularly effective at identifying the paths of attacking and retreating columns of Iraqi armor. Intelligence units made good use of the information, relying on it to establish the precise location of Iraqi units during the battle for the town of Al Kahafji.[59]

Other missions occurred throughout the 1990s. In 1995, during Operation Joint Endeavor, Joint STARS supported NATO peacekeeping efforts in Bosnia. Flying from Rhein-Main AB, Germany, a testbed E–8A and a preproduction E–8C aircraft monitored troop movements on the ground as part of the Dayton Peace Accords. The mission also encompassed surveillance of Bosnian airspace, which involved linking up and communicating with F–16s flying from Aviano AB, Italy.[60] Subsequent missions included Allied Force[61] in 1999 and, most recently, Iraqi Freedom.[62]

The conspicuous place AWACS and Joint STARS occupied in Air Force operations throughout the 1990s and their roles in Enduring Freedom and Iraqi Freedom speaks to their importance as airborne remote sensors. Both systems have earned niches in current and future Air Force missions, but neither attracts the interest and publicity that UAVs, the most recent addition to Air Force airborne remote sensing platforms, now do (Figure 7). Interest no doubt lies, at least in part, with the novelty of the technology, whose practicality and utility for airborne remote sensing only began to emerge during the 1990s. In 1996 the Air Force activated three UAV squadrons of Predators at Indian Springs Air Force Auxiliary Field, Nevada.[63] Equipped with electro-optical and infrared sensors, Predator can fly 24-hour missions 5,000 miles from its home base.[64] Global Hawk, another Air Force UAV, began flying in 1998.[65]

Despite their relatively recent appearance, UAVs have found a range of operational uses. They collected intelligence on Serbian air activity during Deny Flight and played an important role in the air campaign during Enduring Freedom and operations in

Figure 7. Unmanned Aerial Vehicle (UAV).

Afghanistan after the 11 September 2001 attacks. Equipped with sensors, they collected ISR information and battle damage assessment (BDA) data, without endangering a pilot. Their loiter time over target proved another asset. During Iraqi Freedom, R–Q1 Predator UAVs provided aerial reconnaissance and targeted Iraqi radar-guided antiaircraft systems.[66] Predators also performed electronic countermeasures.[67] The other Air Force UAV, the R–4A Global Hawk, saw action too.[68] UAVs, working with AWACS and Joint STARS, gave remote sensing yet more capability. That they would become more and more important to airborne remote sensing seems inevitable.

More significant than any particular airborne sensing platform, though, was the overall capability achieved. Nowhere was this more dramatically demonstrated than in Enduring Freedom and Iraqi Freedom. Both operations showed how vast amounts of data, principally from ISR sensors, could be collected from,

literally, around the globe, exploited in the United States, and then applied to combat operations thousands of miles away. General Joseph Stein, Air Combat Command Director of Operations, has noted how within scant minutes, products were in the hands of operators at the Combined Air Operations Center (CAOC) in Southwest Asia and fed to strike platforms for attack.[69] In short, these airborne sensing systems possessed reachback capability, that is, the means to rapidly access information and specialized personnel skills in the United States during war. This not only reduced the size of the deployment required but, more importantly, compressed the so-called kill chain, the process of finding, fixing, targeting, engaging, and assessing targets.

In one instance, target analysts in the United States, reviewing live Predator videos, found Iraqi tanks hidden in tree lines and relayed the information to the CAOC, which directed their destruction minutes later.[70] On another occasion, imagery analysts in the United States, reviewing live Global Hawk imaging data of the battle area, discovered a SAM site in two minutes. The information, which went to the CAOC, allowed B–2s to destroy the target. Later, the same Global Hawk provided BDA of the target area. The whole process, from initial discovery of the SAM site, to its destruction, to the BDA report that went worldwide, took but a scant 80 minutes. Other operations occurred even faster. Modification to the Predator video, for example, permitted live transmissions to C–130 gunships attacking ground targets.[71]

Such airborne remote sensing technology provides the unprecedented capability to enter, as it were, the enemy's decision loop and destroy his forces quickly, before they can be placed in battle. Furthermore, General Stein predicts, future advances will undoubtedly enable us to compress that kill chain even further.[72]

CONCLUSION

The last half century has witnessed a steady improvement in airborne radar remote sensing. Better vacuum tubes boosted power; solid-state components reduced size and improved reliability; and computers permitted the first practical phased array antennas. These improvements in turn created new missions for radar and enhanced airborne ISR generally. Yet important parts of the surveillance mission remained unmet. Developments of other military technologies, in particular, radar-absorbing aircraft surfaces and precision airborne munitions capable of striking radar transmitters at unprecedented distances, revealed the need for additional measures, and Air Force laboratories set about fashioning them. Bistatic radar,[73] photonics, infrared technology, and passive detection techniques, to cite but a few examples, received increasing emphasis for their potential to bolster radar as an airborne remote sensor. Airborne remote sensing, while still heavily dependent on radar, more and more came to rely on a myriad of sensors. Radar continued to hold center stage in the airborne remote sensing mission, but its relative importance declined as other technologies assumed increasingly greater roles (Figure 8).

Viewed through the prism of history, this

Figure 8. An illustration of AWACS operations.

change appears less a break with the past than a logical consequence of technological advance, which the Air Force, to its credit, readily recognized and implemented. Air Force research, for instance, always emphasized that effective use of air power depended on instantaneous (real-time), useful sensor information and that it included more than just radar. The problem was combining information and displaying it in the cockpit, which during the 1950s and 1960s was technically impracticable. But the ideal persisted, drawing increased strength whenever technological advance seemed to bring it closer to realization. This persistence proved fortuitous since, when enabling technologies did materialize, concepts stood ready for application, having been tried and practiced for years, sometimes for decades. Air Force experimentation with, say, long-range navigation techniques in the 1950s for precision bombing and reconnaissance, while it appears primitive today, actually played a critical role in providing the institutional experience needed to exploit satellite communications, the integrated circuit, the laser, and other technologies as they became available.

Similarly, in airborne remote sensing research, long-established principles and concepts helped prepare the way for the introduction of technologies like personal computers and microcircuitry. Application of new technology, moreover, often challenged existing engineering assumptions and areas of research, once considered separate, frequently merging them in response to innovation. Consequently, radar and radio became integral to communications satellite research; laser and solid-state electronics to munitions research; and the computer, arguably, to all research. Absent the accumulated institutional experience of its laboratories in researching and developing and experimenting with aviation and related technologies, it is difficult to imagine the Air Force managing technological change, imperfect though it sometimes was, effectively. Air Force technological innovation did not occur in a vacuum; it took root in institutional ground nurtured by decades of R&D, R&D that ran the gamut from papers published in research journals and symposia to technology demonstrations and experiments at the laboratory workbench (Figure 9).

The course of technological change, especially for technology advancing as rapidly as airborne remote sensing, deifies prediction. Nevertheless, certain trends stand out, the most obvious being the continuing growth of sensor capabilities and the advantage they bring to air combat operations. Not as obvious perhaps, but equally significant, are burgeoning capabilities for receiving airborne remote sensing information and putting it to timely use, primarily for decision making. The latter may still lag the

former in clearly manifest utility, but the gap seems to have closed. Indeed, airborne remote sensing may have reached a stage in which consolidation and implementation of existing technology more nearly become the norm. Innovations will continue, but emphasis may shift to new stratagems and operational doctrines. If so, the next round of advancement in airborne remote sensing is as likely to come from the hand of the operator or strategic planner as from the researcher or technologist.

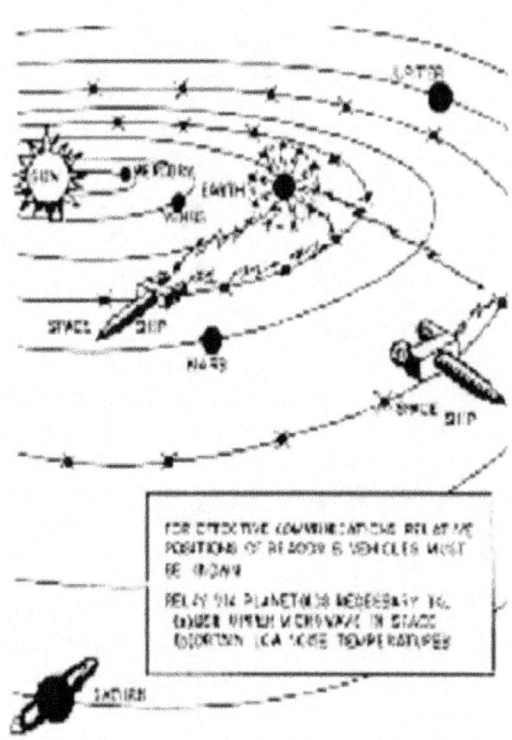

Figure 9. An illustration of COMSAT research.

Pilots in Extreme Environments:
Enforcing U.S. Foreign Policy from the Edge of Space

Major General Robert F. Behler, USAF, Ret.

Russell R. Burton

George C. Mohr

William J. Sears

Billy E. Welch

Abstract

The USAF SR–71 Blackbird is a superb example of the application of science and technology in service to the nation. The far-sighted and often heroic efforts of human system researchers laid the foundation for life-support technologies that allowed pilots to operate in extreme environments. As a result, the SR–71, since its development in the 1960s, has taken pilots and reconnaissance systems officers to the edge of space and back while providing the nation's top decision makers awesome capabilities in a single platform to demonstrate national power.

From 80,000 feet and at Mach 3, the Blackbird reliably provided our leaders critically important national intelligence from an evolving array of the most advanced reconnaissance sensors that the U.S. science and technology community could provide. Underlying this impressive capability are the steady developments in human systems research, which led to ever-improving capabilities for crew members to fly higher, faster, and longer.

The role of the Blackbird in the 1984 MiG21 crisis is a case study in the nation's use of intelligence. In October 1984, intelligence reported that Soviet MiG fighters were crated and being shipped on a Bulgarian freighter. Three SR–71 missions, piloted by the first author, provided decision makers critical intelligence on the location and destination of the possible MiGs while placing our Soviet and Sandinista adversaries on notice that we were watching closely and were willing to act to enforce the Monroe Doctrine.

INTRODUCTION

Late in October of 1984, I was tasked to fly an intelligence reconnaissance mission over Nicaragua. As a result of rising tensions between the United States and Communist regimes developing around the world, I would be an integral player in the demilitarization of Nicaragua (Figure 1). My mission was to fly the fastest, most physically demanding plane the world has ever known to gather vital data on the Soviet Union's operations in and around Nicaragua. Based on previously gathered intelligence data, the U.S. government believed these two countries were exchanging advanced weapon systems to rival neighboring countries.

As an SR–71 Blackbird pilot, I spent many fast hours in extreme environments, performing high-altitude reconnaissance, while defending America's freedoms from those who would oppose us. The earth is much smaller when you're traveling three times the speed of sound at 80,000 feet just along the edge of space in the world's fastest manned aircraft. The SR–71 was the premier high-altitude reconnaissance aircraft designed to provide the right information at the right time to the right person.

Development of the SR–71 began during the administration of President John F. Kennedy in 1962. President Lyndon B. Johnson announced in late February 1964 the existence of a Blackbird prototype, the first experimental jet aircraft that could maintain sustained flight in excess of 2,000 mph at an altitude above 70,000 feet. The first Blackbird (A–12) pilots donned their newly developed S901 full pressure suits and breathed pure oxygen to protect their bodies from the unbearably low pressure. When the SR–71 was finally developed several years later, highly sensitive sensors were installed making the aircraft more conducive to gathering highly specific intelligence data. Equipment bays in the aircraft carried compressed-length cameras, which used highly advanced optics, allowing the SR–71 to photograph 100,000 square miles of terrain in one hour, a landmass larger than the state of Wyoming. In 1972, the Joint Chiefs of Staff's Joint Reconnaissance Center prioritized the purpose of the SR–71, which was to gather intelligence information for five specific intelligence agencies. This brought about a highly intricate development of the premier intelligence-gathering aircraft in the Department of Defense (DoD), which would be the primary source of information for those agencies.

In early October 1984, intelligence analysts pieced together various bits of information

Figure 1. Maj Gen Robert Behler (author) in front of his SR–71.

that led them to believe Soviet MiG21s were being crated for the Bulgarian ocean freighter Bukuriani for delivery to a destination somewhere in South America. U.S. intelligence-gathering satellites had spotted the freighter in a Soviet seaport in the Black Sea loading supposedly highly sophisticated military defense supplies. Because the crates were so large, it seemed logical that the defense materials being supplied would be jet fighters and MIG21s. These events set the stage for a critical mission made possible because of the unique flight performance of the SR–71 and the physiologic protection systems available to its aircrew. In reality, preparation for this mission began sixty-six years earlier. Over succeeding decades, generations of research scientists probed the limits of human tolerance to the extreme conditions characteristic of the SR–71 flight environment. The basic science contributions from wide-ranging disciplines, when blended together, provided the human systems technology required for the successful completion of this landmark SR–71 mission.

Medical Pioneers: 1903–1930

Shortly after the first powered flight in 1903, scientific interest in the medical aspects of aerial flight grew dramatically. It became obvious very quickly that this extreme environment of aviation was different from the environment on the ground. As the country prepared to move into air operations in World War I, the War Department was keenly aware of the need to improve the fitness and efficiency of military aviators to perform combat operations. In 1917, a Medical Research Board was chartered to investigate all conditions that affect the

efficiency of pilots, to determine the ability of pilots to fly at high altitudes, to develop suitable apparatus for supplying oxygen to pilots at high altitudes, and to consider all matters relating to the physical fitness of pilots. This led to the establishment of the Medical Research Laboratory of the Army Signal Corps on Friday, 19 January 1918, at Hazelhurst Field on the outskirts of Mineola, New York (Figure 2). General Harry G. Armstrong, the second Surgeon General of the U.S. Air Force remarked, "The Air Service Medical Research Laboratory was the first of its kind to be established and its contributions to aviation medicine are incalculable in relation to the saving of lives and equipment. Of equal importance is the fact that this institution was the medium through which aviation medicine in all its ramifications was placed on a sound scientific basis in America." The laboratory's research scientists initially focused on developing pilot selection standards and understanding the human effects of exposure to high altitude.

Even in these early times, aeromedical scientists were well aware that oxygen want (hypoxia) was the pivotal hazard encountered

Figure 2. The home of the original Medical Research Labratory on Huzelhurst Field, NY (1918).

during aerial flight. The effects of hypoxia had been thoroughly investigated by a French physician-scientist, Paul Bert, who performed 670 separate experiments from 1870 to 1878 on the physiologic effects of altered atmospheric pressure. In 1874, Professor Bert subjected two aeronauts, balloon pilot Théodore Sivel and engineer Joseph Croce-Spinelli, to a simulated altitude of 23,000 feet in the low-pressure chamber installed in his laboratory. The aeronauts learned about the use of oxygen to prevent hypoxia. On 15 April 1875, Gaston Tissandier joined the aeronauts as a passenger on a balloon flight that reached an altitude of 28,820 feet before descending on its own accord after all three occupants had lost consciousness. Unfortunately, they had decided not to use the onboard oxygen until it was too late to do so. Tissandier survived; his two companions did not the first reported casualties due to hypoxia.[1]

During World War I combat pilots soon found it necessary to fly above 15,000 feet to avoid lethal ground fire. Shortly thereafter, reports began trickling in about troubling symptoms including headache, loss of muscle strength, dizziness, and extreme fatigue. In addition, unexplained losses of aircraft began to accumulate. The medical authorities recognized the root cause as oxygen want. Accordingly, a major effort was mounted by the Army Air Service to develop an oxygen delivery system for pilots performing aerial combat. In 1918, production of the Clark-Dreyer Oxygen System, consisting of an automatic regulator and a leather and rubber mask, got underway. The war ended before the new oxygen system could be installed in other than a small fraction of the Army Air Service's combat aircraft.[2]

Shortly before the end of the war, a young chemical engineer, Lieutenant Harold Pierce, joined the Air Service Medical Research Laboratory after he completed a teaching fellowship in physiology at Harvard University. In 1919 he designed a second-generation altitude chamber fabricated by the Lancaster Iron Works (Figure 3). The chamber, insulated with cork and equipped with a refrigeration unit, enabled scientists to study human response to combined cold stress, reduced atmospheric pressure, and oxygen want that occur at altitude. During unmanned tests, the chamber reached an equivalent height of 75,000 feet at a temperature of minus 31 degrees Fahrenheit. This new chamber was the most advanced piece of experimental equipment in the world. Designed specifically for high-altitude research, it obviously outstripped the technological capabilities of aircraft of the time. Using this facility, medical pioneers at Hazelhurst built a scientific foundation for the development

Figure 3. Mineola Low Pressure Chamber.

of modern-day protective flying equipment that would enable aircrew to fly higher, faster, and longer with each passing decade.

In November 1919, the Air Service Laboratory moved to nearby Mitchel Field and on 18 November 1922, was subsequently redesignated the School of Aviation Medicine. Mitchel Field was named in honor of a former New York City mayor, John Purroy Mitchel, who was killed while training for the Air Service in Louisiana.[3] Before the war ended, Mitchel Field served as a major training base for the rapidly expanding Air Service and proved to be an ideal home for the new School of Aviation Medicine. Four years later, in the summer of 1926, following the rapid postwar drawdown of the Air Service, the War Department decided to move the School of Aviation Medicine to Brooks Field, Texas, collocated with the flying training program still active on that air base. The school's research program was redirected to focus on understanding the practical requirements for the care and selection of the flyer. The Mineola chamber was declared surplus and subsequently shipped to the Equipment Branch at Wright Field, Ohio. This decision was based on the school commander's annual report that declared: "There is reason to believe that the facts of physiology which have been so extensively investigated during the past six years are far in advance of the immediate requirements for the Air Service."[4] This conclusion proved to be false, but it serendipitously set the stage for a major resurgence in scientific investigation of the physiologic requirements for flying at high altitudes.

The Genesis of Modern Air Warfare: 1930–1950

After Charles Lindbergh's solo, trans-Atlantic flight and Jimmy Doolittle's successful all-instrument flight, the fledgling Army Air Corps recognized the potential of airpower in a world heading for global war. Shortly thereafter, in the fall of 1929, a young physician named Harry George Armstrong graduated from the School of Aviation Medicine and decided his future also lay with the rapid growth of military aviation. During his first assignment as the flight surgeon for the famed First Pursuit Group at Selfridge Field, Michigan, Dr. Armstrong discovered for himself how inadequate was the military pilot's protective equipment (Figure 4). While flying in a P–16 open-cockpit pursuit plane from Minneapolis, Minnesota, to Chicago, Illinois, on a winter day in 1934, he discovered his flight clothing provided little protection against the elements. Exposed to a minus 40 degrees Fahrenheit air temperature, he suffered severe frostbite, and his aviator's goggles frosted over, obscuring his vision. Moreover, no oxygen mask was available to compensate for altitude effects. Following his return to Selfridge, he thought about the obvious physiologic threat to combat effectiveness and decided to write a letter to the Air Surgeon in Washington recounting his experiences. He concluded his letter with a strong recommendation that the Air Corps Research and Development Center at Wright Field address the deficiencies in protective flying equipment immediately. As a result, Captain (Dr.) Armstrong was "rewarded" with an assignment to the Equipment Branch of the Engineering Section at Wright Field to serve as an aeromedical advisor. So began the distinguished career of a prolific aeromedical scientist whose pioneering research led to the development of progressively more effective protective flying

Figure 4. Captain Harry Armstrong, Flight Surgeon, First Pursuit Group, Selfridge Field, MI, 1913-1934.

equipment essential for the SR–71 aircrew to accomplish their mission objectives.

In 1935, Armstrong established the Physiological Research Unit as a branch in the Equipment Section of the Materiel Division at Wright Field. He discovered the Mineola chamber sitting idle, covered with dust in a storage room in the basement of the Equipment Branch laboratory building. He had the chamber refurbished and used it for two years to conduct research on the physiologic effects of altitude. During this period, the Air Service had an intense interest in developing a capability for long-range bombardment, which

would again redefine the meaning of extreme environments. The Equipment Branch was assigned responsibility for development of a sealed pressure cabin for high-flying bomber aircraft. Dr. Armstrong was tasked to define the physiologic requirements for inclusion in a sealed cabin aircraft specification.[5]

To meet this challenge, Captain Armstrong recruited several talented scientists from the academic community. One was J. William (Bill) Heim, Ph.D., about to complete postgraduate training in physiology at Harvard University. Dr. Heim accepted Captain Armstrong's invitation to join the Wright Field Laboratory and remained to serve with distinction for more than thirty-one years. Bill Heim, after his retirement, reflected on his perceptions about Captain Armstrong, writing in his memoirs,

I should like to include some observations of this remarkable man who had such a profound influence on the future of aerospace medicine. I was soon to learn that Armstrong was a dedicated medical officer, soft spoken, with great personal charm and possessing a strong but not uncritical loyalty to the service. I was always amazed by the quiet, relaxed, yet self-assured manner in which he carried out his activities. No thrashing about, no hurried pace, no long over-time hours, yet everything he did seemed to count. As an outstanding characteristic, he appeared to be thoroughly enjoying everything he did and his enthusiasm was infectious. An almost undetected talent was his ability as an entrepreneur par excellence. With his disarming and convincing manner, he was a master of the soft sell; one found it extremely difficult to say no. Armstrong with his creative mind was a penetrating observer and a superb pragmatist, with the almost uncanny ability to isolate the core of a problem, perform a minimum of critical experiments, and apply the results to a practical solution.[6]

Dr. Armstrong and his scientific team successfully developed the design requirements for pressurized cabins (Figure 5). The prototype system was incorporated into the

Figure 5. Dr. Armstrong (seated) at work with his new altitude chamber (1937).

XC35 aircraft, delivered by the Lockheed Corporation to Wright Field for flight-testing in the spring of 1937. Sealed-cabin technology was subsequently widely applied for inclusion in future commercial passenger aircraft and advanced Air Corps bomber aircraft beginning with the B29 Stratofortress.

Recognizing that in addition to the effects of altitude, pilots were exposed to substantially increased G-forces during aerial maneuvers. Dr. Armstrong designed and had installed in a vacant balloon hangar on Wright Field the first human centrifuge to be used in the United States. Though exceedingly unsophisticated by modern standards, this novel research

tool fabricated in the Equipment Branch machine shops served well. Dr. Armstrong performed extensive studies of the effects of accelerative forces on blood pressure, first using goats and finally humans.

Dr. Armstrong's research contributions encompassed virtually all aspects of aerospace medicine. The majority of his investigations were the first of their kind to be carried out anywhere in the world. An abbreviated list of the aeromedical problems he investigated during the six years he was director of the laboratory include the following:[7]

- Oxygen want (hypoxia) and requirements for supplemental oxygen
- Reduced atmospheric pressure effects on the middle ear, nasal sinuses and dental fillings
- Explosive decompression, the risk of gas bubbles forming in the body and pre-breathing requirements
- High altitude flight stresses including cold exposure loss of body fluids and flying fatigue
- High positive and negative acceleration effects on blood pressure and vision
- Pilot vertigo, airsickness and spatial disorientation
- Toxic hazards in the cockpit including carbon monoxide and radioactive materials

Dr. Armstrong perceived from the outset that the human element was one of the most important factors in aircraft system design; yet design engineers of his time paid little attention to the pilot's needs for protection against the harsh environments encountered in flight. His goal was to develop robust, protection system design criteria backed up by rigorous scientific data, anticipating future advances in aircraft system development. This vision of Dr. Armstrong was strengthened and expanded by the pioneers who succeeded him. Another young physician, Dr. Otis O. Benson (then a captain) became the second chief of the laboratory. Under his direction, the Aeromedical Research Unit was withdrawn from the Equipment Laboratory and made a separate laboratory with three units of its own (Physiological, Biophysics, and Clinical Research). The Aeromedical Laboratory moved from its overcrowded quarters to a new building on Wright Field and was joined by its sister organization, the School of Aviation Medicine from Randolph AFB. Dr. Benson organized a research program for the laboratory that persisted throughout World War II. He staffed the laboratory with nationally known scientists, drawing significantly upon the contacts he had developed earlier in his training with the Mayo Clinic and the Harvard Fatigue Laboratory. Collaborating with the Mayo Clinic and other researchers, he established the human centrifuge unit that contributed to the development of the anti-G suit. Before World War II, he recognized the need for a radically different method of supplying oxygen to aircrews during high-altitude bombing. Under his leadership, the diluter-demand oxygen system was designed and perfected.[8]

The Demand for Advanced Technology: 1950–1985

Following World War II, Air Force aeromedical scientists at both the School of Aviation Medicine and the Aeromedical Research Laboratory continued to expand the scientific knowledge needed for new systems.

From this living database came the criteria for protection systems and cockpit designs to enable aircrew to perform safely at extreme altitudes and supersonic speeds, in high and low ambient temperatures, exposed to high maneuvering acceleration forces, intense noise, and vibration, and fatiguing flight durations. As in its earliest days, the aeromedical scientist of the Air Force laboratories also continued to support the flyer in increasingly extreme environments and missions.

With the advent of jet aircraft and the need to escape under adverse conditions, Colonel (Dr.) John Stapp led the way in his pioneering impact and deceleration research to define the limits of human tolerance. He recognized the Air Force would continue to fly higher and faster until it all but shattered the barriers of physical forces, but human limitations would persist unchanged with each new generation of aircraft. Early in his quest to define human impact limits, he earned the title The Fastest Man Aliveî when, on 10 December 1954, he rode the Sonic Wind I rocket sled, attaining a maximum speed of 639 mph in 5 seconds and decelerating in 1.25 seconds, sustaining a peak stopping-force of more than 40 Gs. His body, for brief moments, weighed 6,800 pounds (Figures 6, 7). The windblast and deceleration forces at his top speed were roughly equivalent to ejecting in an open seat at three times the speed of sound and from an altitude between 55,000 and 60,000 feet. Escape from an aircraft during flight at high altitude and supersonic speed exposes the pilot to a variety of potentially lethal events. First the ejection force required to clear the aircraft empennage can cause crushing injury to the spinal column. Then, the sudden exposure to windblast and wind-drag deceleration can cause the limbs to flail and induce dangerous tumbling and spinning. In addition, the pilot ejecting at high altitude needs to be provided supplemental oxygen and be protected against exposure to intense cold and dangerously low barometric pressure. When Colonel Stapp began investigating these risk factors at Holloman AFB in the 1950s, aircraft escape systems were either inadequate or of unproven worth for aircraft flying faster than Mach 1 or at altitudes above 45,000 feet. Over the ensuing three years, Colonel Stapp personally made twenty-seven of the seventy-three manned sled tests conducted as part of the deceleration project (Figure 8). The research data obtained from these groundbreaking studies, and from follow-on work using the Daisy Decelerator (a specially designed human impact simulator), defined human tolerance to

Figure 7. Dr. Stapp Windblast Exposure.

Figure 6. Dr. Stapp riding Sonic Wind I Rocket Sled.

windblast and a broad range of impact forces in all planes of body orientation. This data provided the designers the means to develop successive generations of highly capable escape systems.[9] During the 1970s and 1980s other aeromedical scientists developed mathematical

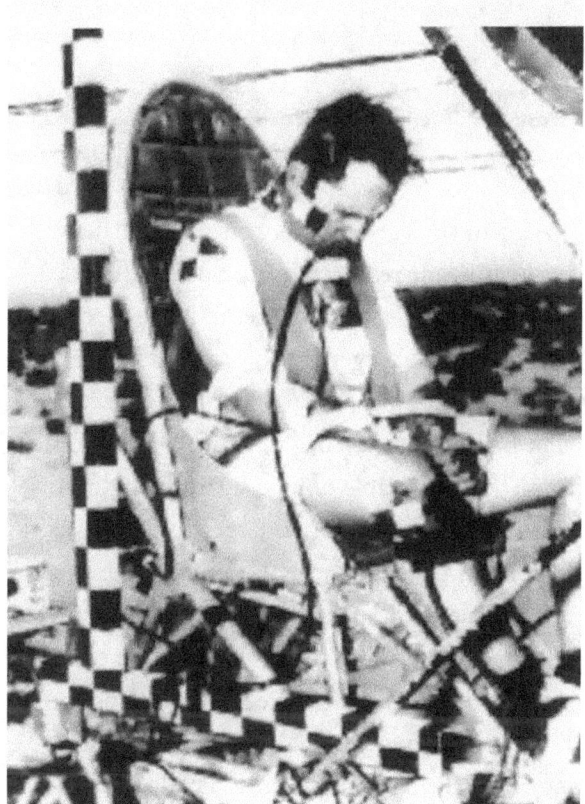

Figure 8. Dr. Stapp Deceleration Test.

models to support the system design process; however, it was the Air Force scientist, Colonel John Stapp, who forged the way.

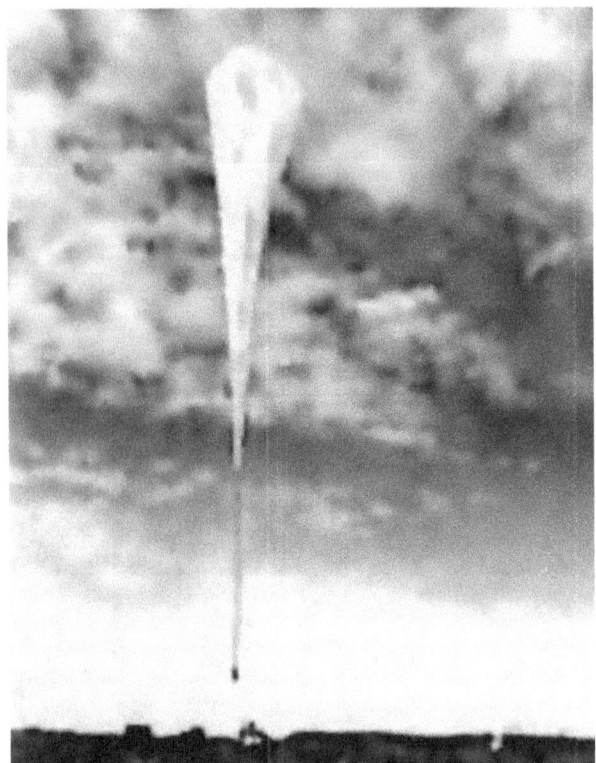

Figure 9. Beginning the ascent in the Excelsior III gondola.

Also in the early part of this period, there was significant research in support of full and partial pressure suit assemblies to meet high altitude emergency requirements. Stapp recruited Captain Joe Kittinger for Project Man High, a project begun in 1955 that would use balloons capable of high-altitude flight and a pressurized gondola (the basket or capsule suspended from the balloon) to study cosmic rays and determine if humans were physically and psychologically capable of extended travel at space-like altitude (above 99 percent of the Earth's atmosphere). The Air Force had determined that a high-altitude balloon flight was the best way to conduct these studies because the periods that aircraft could remain at these altitudes would be too short to provide useful data. Using a two million cubic foot (56,634 cubic meter), 172.6-foot (52.6-meter) diameter balloon and a cramped aluminum alloy capsule manufactured by Winzen Research of Minneapolis, Kittinger made the first Man High ascent in June 1957, remaining aloft for almost seven hours and climbing to 96,000 feet (29,261 meters). Two additional Man High flights were made, one by Major Dave Simons to an altitude of 101,516 feet (a world altitude record at that time) and the other by Lieutenant Clifton McClure to 90,000 feet. Subsequent flights to test equipment and explore escape from high-altitude platforms were also flown in balloons, but with open gondolas. In 1960, Captain Joe Kittinger, floated to 102,800 feet (31,333 meters) in Excelsior III, an open gondola adorned with a paper license plate that his five-year-old son had cut out of a cereal box (Figure 9). Protected against the subzero temperatures by layers of clothes and a pressure suit he experienced air temperatures as low as minus 94 degrees Fahrenheit (minus 70 degrees Celsius) and loaded down with gear that nearly doubled his weight, he climbed to his maximum

Figure 10. Captain Kittinger photographed by an automatic camera as he prepared to jump from 102,800 ft.

Figure 11. Captain Kittinger stepts into space.

altitude in one hour and 31 minutes, even though at 43,000 feet (13,106 meters) he began experiencing severe pain in his right hand that occurred as a result of a failure in his pressure glove and could have scrubbed the mission. He remained at peak altitude for about 12 minutes before he stepped from his gondola into the darkness of space (Figures 10, 11). After falling for 13 seconds, his 6-foot (1.8-meter) canopy parachute opened and stabilized his fall, preventing the flat spin that could have killed him. Only 4 minutes and 36 seconds were needed to bring him down to about 17,500 feet (5,334 meters), where his regular 28-foot (8.5-meter) parachute opened, allowing him to float for the remainder of the way to Earth. His descent set the record for the longest parachute freefall, a record that stands today.[9]

In the 1960s and 1970s, the appearance of new operational requirements and new technologies to make them possible simultaneously led to another round of the definition of extreme environments. Flying higher, faster, longer, more maneuverably, and with increased workload, this all combined to stimulate the development of new scientific data to expand the envelope "and create new technologies to support the aircrew in their mission of To Fly and To Fight." A comprehensive list of aircraft, medical, and protective equipment issues that must be considered for high-altitude flight is shown in Table 1.

Physiological Issues of High Altitude Flight

In high-altitude flight, including space flight, a structural failure in a pressurized cabin or loss of cabin pressure control would be catastrophic without protection for the crew. Physiological effects of rapid decompression include acute hypoxia, effects on the gas-containing cavities of the body, decompression sickness, ebullism (vaporization of body fluids), and thermal exposure. Less rapid but equally debilitating effects of unpressurized flight at high altitude include hyperventilation, fatigue, reduction in effective circulating blood volume, and fainting associated with pressure breathing. Also, acceleration forces during high-speed egress can have a profound consequence on the skeletal structure and the cardiovascular system. All can have potentially grave effects upon aircrew performance and mission effectiveness, and all are avoidable by the employment of appropriate life-support equipment and adequate training. The SR–71 pressure suit maintains the crewmember at an equivalent altitude of 35,000 feet breathing 100 percent oxygen (equivalent to being at sea level and breathing air), thereby preventing most of these problems.[10, 11]

Hypoxia

Altitude hypoxia results when the oxygen partial pressure in the lungs falls below that comparable to sea level, but it is insignificant until the alveolar oxygen tension falls below a 10,000-foot equivalent. At 10,000 feet, the reduced ability to learn new tasks can be measured; consequently, 10,000 feet is used as the altitude that supplemental oxygen is considered necessary. As the partial pressure of oxygen in the inspired air continues to drop, the signs and symptoms of hypoxia become more evident and include loss of peripheral vision, skin sensations (numbness, tingling, or hot and cold sensations), cyanosis, euphoria, and eventually unconsciousness at higher altitudes. Up to an altitude of 34,000 feet, increasing the percentage of oxygen to 100 percent allows a sea-level oxygen equivalent. Above 40,000 feet, breathing 100 percent oxygen without additional pressure is not sufficient for efficient aircrew performance. Positive pressure breathing is required and is accomplished by use of an oxygen system that delivers 100 percent oxygen at greater than ambient pressures. Without higher than tolerable pressure breathing, even a short exposure to altitudes higher than 50,000 feet leads rapidly to unconsciousness. Thus, all aircrew must wear a partial or full pressure suit above this level.[11, 12]

Mechanical Effects

During a cabin depressurization, gases trapped within the intestinal tract, nasal sinuses, middle ear, and lung will expand. The magnitude of the effect on the gas-containing cavities of the body is directly proportional to the range and rate of change of pressure. Serious consequences result when an occlusion or partial occlusion occurs between a gas-containing cavity and the environment.[11, 12]

Decompression Sickness

Body tissues contain dissolved gases, principally nitrogen, in equilibrium with ambient atmospheric pressure. When ambient pressure is reduced, nitrogen bubbles form in body tissues. If the drop in pressure is not too great or too fast, bubbles evolved in the tissues are safely carried by the vascular system to the lungs where the evolved nitrogen is eliminated. Prolonged exposure to altitudes in excess of 25,000 feet (occasionally between 20,000 and 25,000 feet) may lead to one or more of the symptoms of decompression sickness, that is, bends, chokes, and circulatory and neurological disturbances Recent research has established the need for increasing the pressure differential (from 5 to 7 psi) in future aircraft that may fly at these higher altitudes.[13] An increased variable-pressure differential has recently been suggested by Air Force researchers to provide less risk of decompression sickness resulting from potential of prolonged exposure to cabin altitudes in excess of   ,000 feet.[14]

Ebullism

When the total barometric pressure is less than the vapor pressure of tissue fluid at body temperature (47 mm Hg), vaporization of the body fluids occurs. This occurs in the nonpressurized portions of the body at altitudes above 63,000 feet. However, exposure of peripheral regions of the body, for example, the hands, to pressures less than the vapor pressure of the tissue fluids leads to vaporization of these fluids with little or no impairment of performance. When combined with low environmental temperatures, the evaporative cooling associated with vaporization may accelerate freezing/drying of exposed tissues. When wearing only a mask for short exposures above 63,000 feet, vision is affected as a result of tearing and blinking during positive pressure breathing, effectively blinding the crewmember.[14, 15]

Thermal Extremes

Low temperatures following the loss of cabin pressure at high altitude can cause impaired function and eventual tissue damage to exposed regions of the body or, in longer duration exposures, a drop in core temperature leading to progressively impaired performance followed by unconsciousness and eventually death. An aircrew member wearing normal flying clothing with mask and gloves will not suffer any serious damage during a short exposure (5 minutes) to the lowest temperature conditions encountered at high altitude. Exposure beyond this time will lead to more severe peripheral cold injury unless appropriate clothing/heating garments are worn.[2, 3] The garment must also protect against the heat where the temperatures on the outer surfaces of the SR–71 at cruise approach 560 degrees Fahrenheit as well as provide heat protection during the thermal pulse of ejection at Mach 3.[16]

Pressure Suit Development

The first recorded suggestion for the use of pressure suits was by J.S. Haldane in 1920, who stated:

If it were required to go much above 40,000, and to a barometric pressure below 130 mm Hg, it would be necessary to enclose the airman in an air-tight dress, somewhat similar to a diving dress, but capable of resisting an internal pressure of say 130 mm of mercury. This dress would be so arranged that even in a complete vacuum the contained oxygen would still have a pressure of 130 mm Hg. There would then be no physiological limit to the height attainable.

Military application of pressure suits was limited in the early years, and efforts involving high-altitude protection were generally left to adventurers and their scientific advisers. Early pressure-suit development flourished as a result of both aviation and balloon contests. Over the years a multitude of developmental and operational pressure protection systems have been produced. Most of the developmental and production pressure suits naturally evolved as attempts to provide a more comfortable, lightweight, and functional protective system that conformed to the requirements of specific operational conditions, for example, from short-term exposure to altitudes above 50,000 feet to moon walks in a vacuum. Full pressure suits (protective ensembles with associated regulators, oxygen systems, and ancillary hardware that completely enclose the aircrew member) have been shown to provide long-term protection against many of the effects of high-altitude exposure, but their acceptance has been limited, except for high-altitude reconnaissance or flight-test missions. Their use often involves restrictions to the pilot's visibility, mobility, and dexterity and tends to reduce mission effectiveness, as when a visor in front of the eyes makes night refueling difficult, even dangerous. Additionally, the visor acts as a condensing lens and becomes especially disturbing when the sun is in the forward field of vision.

The first full pressure suit was developed by an English firm for American balloonist Mark Ridge in 1933. The suit was taken to 84,000 feet with the body pressurized to 36,500 feet. This suit was used to break two world records in 1935. After several attempts to develop a suit that was reasonably comfortable, B.F. Goodrich built a suit of double-ply rubberized parachute fabric for Wiley Post in 1934. Components included pigskin gloves, rubber boots, and an aluminum helmet. The suit was pressurized to 7 psi, and ten flights were conducted before Post's death in 1935. Several other countries, most notably the USSR, England, Germany, France, and Italy, developed full pressure suits during this period. Most of the emphasis in the newly formed U.S. Air Force in 1947, however, was directed toward partial pressure suits, suits that partially enclose the body and apply mechanical counterpressure, which generally provides shorter term protection to the effects of high-altitude exposure. On the other hand, the U.S. Navy placed their emphasis on omni-environmental, full pressure suits to combine altitude and immersion protection.

From 1943 through 1948, the Air Force's Dr. James Henry and others from the University of Southern California designed the capstan partial pressure suit and exposed subjects to 80,000 feet

for varying lengths of time. The David Clark Company subsequently developed Dr. Henry's original capstan partial pressure suit and produced the first operational models in custom sizes for early rocket-powered Xplane test pilots, for example, Yeager et al. They produced the T1 capstan pressure suit in standardized sizes made of nylon cotton twill. It was chamber-tested to 106,000 feet and subsequently flown in a variety of high-altitude aircraft. The T1 capstan suit (5-to-1 capstan to suit ratio) incorporated an anti-G suit, had no chest bladder, and was made in twelve standardized sizes for fighter aircraft. This was followed by several modified capstan suits, for example, the MC1, MB1, MC3(A), MC4(A), CSU2P, and S100, that incorporated standard sizing, chest bladders for easier breathing at extreme altitudes, and looser fit for more comfort, as well as anti-G suits for fighter aircraft. Capstan partial pressure suits were adopted for the original U2 pilots and continued in use until the last original U2C model was retired from service in 1989. A bladder-type partial pressure suit produced by David Clark, the CSU4/P, was used by Colonel Joe Kittinger in the stratosphere jump. Interestingly, a variant of the CSU4/P, the 1032 Launch Entry Suit, was produced by the David Clark Company many years later for National Aeronautics and Space Administration (NASA) astronauts

The B.F. Goodrich Omni-Environmental Full Pressure Suit was developed in 1948 by the Navy. Suits had earlier been developed by Goodrich for the Doolittle mission in 1942. In 1951, an entirely new full pressure suit was produced by the David Clark Company for D5582 Douglas Skyrocket test pilots. It was first flown by Scott Crossfield for his record-

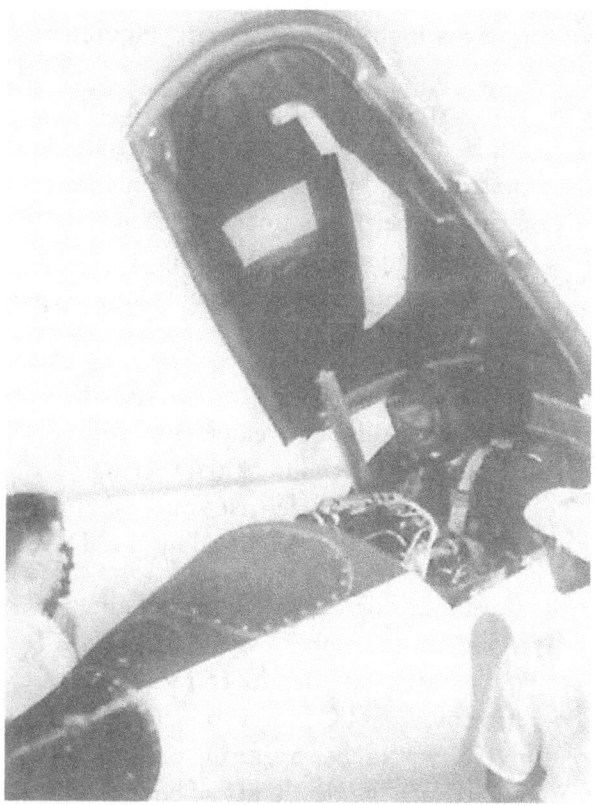

Figure 12. Navy test pilot, Marion Carl, in D–558–2 pressure suit.

breaking Mach 2 flight in the D5582 and later by Navy test pilot Marion Carl for an 85,000-foot altitude record flight (Figure 12). This suit became the forerunner of the Air Force's X15 full pressure suits produced by David Clark Company for Scott Crossfield, NACA (Figure 13). A custom-modified U.S. Navy Mark IV Series full pressure suit produced

Figure 13 Test piolt, Scott Crossfield, in front of his X–15.

by B.F. Goodrich was also developed as a backup emergency system for intravehicular activity for the Mercury program. In the early 1960s, the David Clark Company produced the Air Force's first standardized full pressure suit, the A/P 22S2. Numerous other models (the 4, 6, and 6A) followed over the years.

Except for the space suits produced by ILC Dover for NASA (for the Apollo, Lunar, and Skylab programs), most of the full pressure suits from the 1950s on were produced by the David Clark Company, including NASA, Gemini, and Apollo Block 1 spacesuits. The MC2 full pressure suit was developed with an integrated parachute harness and was first used in the X15 aircraft.

The first high-altitude S901 series full pressure suits were produced by the David Clark Company in early 1960 specifically for the Blackbird (Figure 14). These suits went

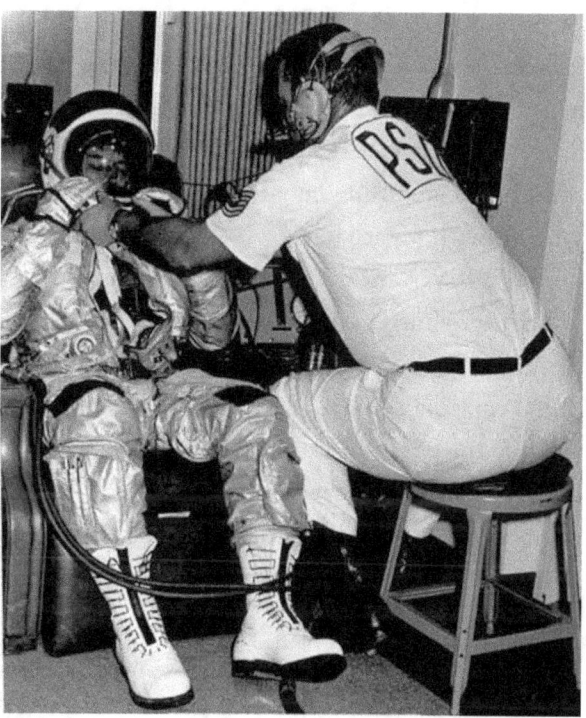

Figure 14. SR–71 pressure suit cira 1968.

through continued design changes through the A12, YF12A, and SR–71 programs. It was a full pressure suit with integrated subsystems, parachute harness, automatic flotation system, urine collection device, redundant pressure control and breathing system, thermal protective garment, custom sizing plus twelve standard sizes, and various models leading to the S901J which was specifically developed for the SR–71 aircraft. In 1977 the S901J was replaced by the S1030 series suit, a full pressure suit with link net and integrated subsystems. This suit incorporated five layers: the first consisted of long cotton underwear, glove liners, a ventilating system, and socks that wick moisture from the skin and provide a means for removing heat buildup; the second was a pressure containment layer; the third, a restraint layer; the fourth, a coverall; and the fifth, a vest that integrated the parachute, survival kit, and flotation system. By the mid-1980s, the S1031 had become the protective suit for the SR–71 as well as the newer U2R aircraft and was later modified to fit female pilots. In the 1990s, the S1034 replaced the S1031 combining integrated life support systems with breathable pressure bladders of Goretex. A variant of the S1034 was adopted by NASA for shuttle astronauts in the early 1990s, designated the S1035 Advanced Crew Escape Suit, replacing the S1032 launch/entry suit.[17]

Over the past three decades, several other suits have been prototyped and tested by U.S. Air Force research laboratories at Wright-Patterson and Brooks AFBs in cooperation with the Life Support System Program Office to provide protection from newer threats as well as to test innovative concepts that potentially would provide greater safety, mobility and comfort for the aircrew of high-altitude aircraft (Figure 15):

- Prototype High-Altitude Flying Outfit (PHAFO): A 1979 prototype partial pressure suit by David Clark to integrate altitude, thermal, immersion, chemical defense, and anti-G protection. It incorporated a nonconformal (Dome Type) full pressure helmet with oxygen mask.

- High-Altitude Flying Outfit (HAFO): A 1979 prototype developmental full pressure suit with integrated thermal/ pressure/chemical defense/immersion and anti-G protection, produced by ILC Dover.

- Advanced High-Altitude Flight Suit (AHAFS): A high-pressure (56 psi) full pressure suit developed for the Air Force by ILC Dover to increase mobility at higher operating pressures and reduce the possibility of bends at the higher cabin altitudes.

- Tactical Life Support System (TLSS): Developed by, among others, the USAF School of Aerospace Medicine, Life Support System Program Office, and Boeing/Gentex to provide short-term protection from 60,000 feet. It incorporated many new features combining a modular high-pressure mask, vest, anti-G suit ensemble integrated to provide Pressure Breathing for G (PBG) for high G-maneuvers, and Pressure Breathing for Altitude (PBA) for altitude, with G-trousers providing four times the breathing pressure from a molecular sieve oxygen-concentration system. As an Advanced Development program, the charter of the TLSS effort was to provide a vehicle to incorporate the laboratory-generated technical advancements into an integrated system to improve aircrew life support. Concepts advanced by TLSS are being used in current life-support systems and are being evaluated for the next generation of life-support equipment. Specifically, TLSS pressure breathing concepts are flying today in Combat Edge and in a pressure breathing system for both altitude and +Gz protection developed for the F22.[18] The United Kingdom, Canada, Sweden, Finland, and France use many variants of similar protective design.

While it is nearly impossible to predict all new threat scenarios that may emerge over the

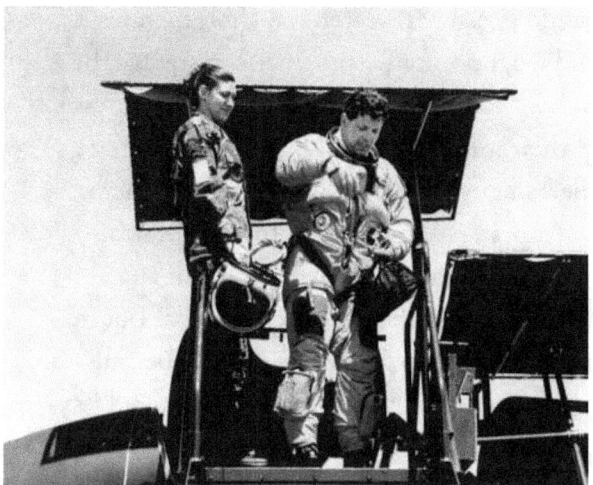

Figure 15. Francis Gary Powers, Jr., suited up for U–2 flight, circa 2000.

next ten to twenty years, attempts to outline anticipated needs for high-altitude systems have been proposed by U.S. Air Force laboratory personnel. For full pressure systems associated with reconnaissance flights, the following apply:

- Protection to altitudes above 80,000 for extended periods (16 hours).
- Ejection/windblast protection.
- Protection against high and low temperatures.
- Protection against nuclear, biological, and chemical (NBC) agents and

provide for selfdon/doff capability in the NBC environment.
- Provide a system that is lightweight, durable, and easily maintained in an NBC environment, to include being decontaminable.
- Provide adequate mobility for accessibility to aircraft and instruments without assistance.
- Eliminate requirement for prebreathing 100 percent oxygen with oxygenation equivalent to breathing air at sea level.
- Allow fluid intake, feeding, urine output, temperature regulation, and increased tactile capabilities.
- Provide a lightweight helmet, optically correct visor with unrestricted vision, visor heating and defogging, anti-reflectance coating, improved spectacle mounting, communications, and improved head mobility.
- Provide anti-drown and anti-suffocation features.
- Provide laser and flash protection.
- Provide coverall with improved hold-down, integrated flotation, parachute harness, fire protection, and ancillary hardware.
- Provide gloves with improved sizing, tactility, and dexterity.
- Provide built-in test features.
- Provide better supportability and reduce training with minimum service life of six years.

The ideal life-support system should not reduce routine flying efficiency. A compromise however is necessary between minimal reduction in performance and adequate protection after loss of cabin pressure or during egress. The success of the crew station pressurization system/configuration and protective clothing and equipment will depend on the correctness of trade studies and evaluations and on the assessment of other requirements such as oxygen regulation systems; head protection from impact and buffeting; eye protection from laser hazards; dazzling glare and nuclear flash; head-up-displays; aircraft escape/evasion; and chemical, biological, and radiation protection.

Applying Science and Technology: The Bukuriani Mission

The real meaning of what we do in our laboratories and why it is vital to our national security becomes clear when you consider the results that can be achieved by these capabilities. One example involved me in an up-close and personalî fashion and, I believe, makes the point about the importance of science and technology.

The circumstances leading up to my mission on 31 October 1984, evolved quickly. The Bulgarian ocean freighter Bukuriani had been continuously tracked during her transatlantic voyage. At times during the several weeks the freighter was tracked, clouds and severe weather blurred the satellite images, and at times U.S. intelligence personnel lost track of the ship. When the weather cleared, they located the freighter in port in Nicaragua where port authority personnel were photographed unloading the suspicious large cargo crates. The U.S. military decided to send reconnaissance aircraft to

verify that this was the same freighter that sailed from Bulgaria carrying sophisticated military aircraft and weaponry.

The U.S. intelligence community, not wanting to be surprised by another Cuban missile crisis or military buildup, tapped a Giant Clipper SR–71 mission to fly from Beale AFB, California, to monitor activities in and around Cuba, focusing on any military activity in Nicaragua. The belief was that the Soviet Union was supplying the Nicaraguan defense ministry so that it could become a dominant communist regime in Central America which would possibly threaten the United States and neighboring Latin American states. The SR–71 crews knew an equally important part of their mission was to show the flag and let the bad guys know we were watching them.

At first, it appeared the Bukuriani was heading through the Mediterranean and straight across the Atlantic for Cuba. But in mid-October, an SR–71 Giant Clipper mission determined that the freighter had altered its course and was sailing around Cape Horn. Its destination, while still unknown, was now

Figure 16. An SR–71 Blackbird at rotation.

thought to be Peru. On October 26th, Major Ron Tabor, my reconnaissance systems officer (RSO), and I were tasked to fly a routine Clipper mission   a normal launch at the crack-of-dawn, California time, refueling thirty minutes after takeoff over Idaho, Mach 3 across the south-central United States, and descending to a KC10 from Barksdale AFB, Louisiana, over the Gulf of Mexico (Figure 16). The weather in the take area "was excellent; therefore we were sure the intelligence data was going to be of the highest quality. We didn't realize it then, but we already had the proof that the Bukuriani was not going to Cuba. It would arrive two weeks later on Wednesday, November 7th, in the Nicaraguan port of Corinto, on the Pacific side. As U.S. voters were reelecting President Reagan for his second term, my crew and another were both in the vault," planning a very unusual Clipper mission. This time the take area was not Cuba; rather, it was Nicaragua.

Because the SR–71 can achieve speeds of over thousands of miles per hour, coupled with the fact that Nicaragua is such a small country, the target area for intelligence reconnaissance looked to be very difficult to approach and photograph. Besides the peculiar target area, this mission was unusual for several reasons. Instead of arriving in the area of interest at about noon, when sun-angles were at their optimum for a photo mission, we were going to arrive just after sunrise in order to wake upî the Sandinistas. Even at altitudes above 80,000 feet, the shock wave from an SR–71 creates window-shaking double-booms on the ground. A Nicaraguan Defense Ministry spokesman later said the sonic booms caused mass panic that morning as Nicaraguan ministers told

the people the United States was mobilizing for an intense war against them. When they heard the "booms" they assumed a military onslaught from the United States was imminent. Residents in Central and South America were accustomed to U.S. reconnaissance missions by both SR–71s and C130 cargo planes, but this time, it was different. They were told to be ready for war with the United States Frightened residents poured into the streets of Managua, Nicaragua's capital city.

Another reason this mission was unusual was the requirement to stay within Nicaraguan airspace. Generally, during night missions, aircraft bank-angles were limited to 30 degrees. On this mission however, because of the speed and small area of interest, we had to bank the SR–71 up to 45 degrees. This type of maneuvering is dangerous because above 80,000 feet at nighttime, visual references to a natural horizon are virtually nonexistent. If we missed the target area, we would have to do a literal U-turn and return to the target area, which at Mach 3 involves hundreds of miles to maneuver and complete. If the aircraft instrumentation and autopilot malfunctioned, the Blackbird would more than likely transition to an uncontrollable, unrecoverable attitude.

The most unusual aspect of this mission was that we were going to launch two SR–71s. The plan was to launch a second, spare SR–71 thirty minutes after the first one departed. If for any reason the primary aircraft aborted, the second aircraft would continue to fly through the take-area. I was scheduled to fly the backup SR–71. The reason we used two aircraft was that this mission had the highest interest level   that of the President. One of President Reagan's issues

107

in his platform for his second election regarded more peaceful foreign diplomacy with the Soviet Union. Because the DoD and the White House believed that Moscow was aggravating U.S. foreign policies with its military support of the Nicaraguan communist government, Reagan's desires for peace settlements with the Soviet Union would be most likely postponed.

At Strategic Air Command Headquarters, a three-star general was the action officer calling the shots on this mission. During the initial planning session, I was designated to fly the backup SR–71, which I didn't expect to fly because the primary aircraft had a very high launch rate. The scheduled launch on October 31st for the first plane was 2:30 AM; my launch was at 3:00 if the primary bird had a problem. As a standard procedure, we started engines forty minutes prior to a launch. Both aircraft went through normal starting and taxiing procedures. By the time I got to the last-chance inspection area at the end of the runway, the primary aircraft was on the runway ready to launch.

Then I heard the primary pilot say "Tower, Aspen 30 request a 180 to taxi off the active and abort the launch." My aircraft was now the only one able to accomplish the mission. The next thing I heard over our special Aspen Control frequency was "good luck Lorenzo" (Lorenzo was my call sign). I asked my RSO "are we ready for this?" His reply to me was a confident "you bet," and then to the tower, "Aspen 31 is ready for departure." We launched at 3 AM and zero seconds. After a great sortie, we returned to Beale at about 9:30 AM. The imagery confirmed the Bukuriani was in port, but the crates had not yet been offloaded. Once more the

Soviet Union and the United States were on opposite sides of an international controversy.

A couple of days later, on November 9th, we were preparing to fly back to Nicaragua, this time in the primary aircraft. International tensions were growing rapidly. Nicaraguan military intelligence personnel claimed a U.S. military ship had entered their waters, posing a threat to their national security. U.S. officials claimed that the ship was more than twelve miles off the coast, not inflicting any threats. Nicaraguan military personnel also fired antiaircraft missiles at a U.S. C–130, which they believed was flying too close to the ground. While the U.S. contended that it was flying a routine mission over South America in international airspace, not threatening the Nicaraguan national security, the Nicaraguan government informed its citizens that the United States was mobilizing to go to war against them and bid its people to prepare to fight U.S. soldiers.

The mission planners were unable to give us a flight profile in advance because they were building it until just a few hours before launch time. My RSO and I finally received the mission profile at 1 AM during breakfast. Fortunately, the only changes were in the take-area. I had to fly at Mach 2.8 because the turning radius was too great for anything faster. Our launch was flawless. Once again the Sandinistas were able to set their alarm clocks at 8:30 in Managua to our thunderous sonic boom. This time they were on the edge of their seats starting to mobilize their citizens for the potential war with the United States. One of their revolutionary commanders said that they were preparing for a Reagan-launched direct and massive war.

By Saturday morning, 10 November 1984, we were exhausted. Two back-to-back SR–71 sorties drained us physically and mentally. This plane was the most intense aircraft to fly, and after just a single mission even the most physically fit have to rest and recuperate. It was time for a break. Since we were on early wakeup for the last two weeks, we were up early and had a normal, relaxing breakfast. When I returned

Figure 17. Enforcing U.S. foreign policy from the edge of space.

home at 11 AM, I found about twenty messages on my answering machine. We were to go back to Nicaragua a third time. My RSO and I were ready, but when we returned to Beale, the State Department declared the MiG21 crisis over.

Though few knew what we were doing, the purpose of the SR–71 missions was to ensure that U.S. foreign policy was enforced in belligerent communist countries. Those who flew these planes were the Air Force's elite pilots and RSOs, as there were fewer SR–71 pilots than there were U.S. astronauts. The world may never know if the Bukuriani was actually carrying crated MiG21 fighters to the Nicaraguan port of Corinto. Officials said privately that the department's objective was to allow the Soviets an opportunity to take the jets home without having to admit they were even there in the

first place. This strategy was believed to be the best way to end this matter peacefull.

EPILOGUE

Military mastery of air and space places the air crewman in the harshest of environments. The in-house aeromedical laboratories are steadfast in their pursuit of basic science, producing new technologies often before a formal operational requirement is endorsed by the user. The lesson to be learned from this historical review is that research consumes a great deal of time, and technology-push is an important research concept that must not be overlooked in the rush to support the user's perceived immediate need.

Forecasting potentially useful new mission capabilities for aircraft entering the operational inventory will sometimes identify future technology challenges not yet validated by the operational community.

The Air Force Research Laboratory's Human Effectiveness Directorate recently initiated an in-house, technology-push project called the Sustained High Altitude Respiratory Protection and Enhanced Design G Ensemble (Sharp Edge).[19] The potential operational need for a new altitude protective system was examined and the physiologic implications of a rapid decompression were addressed. The requirements for an altitude protective system to permit flight above 60,000 feet were outlined, followed by a technology assessment of current life support systems. The Sharp Edge system would be required to interface with current life-support systems, including helmet-mounted displays, chemical and biological defense systems, and oxygen and anti-G ensembles.

These interface and integration requirements were summarized, and the difficulties likely to be encountered were identified Finally, design recommendations were made for the development of the Sharp Edge ensemble. In addition to this effort, a technology watch for newer materials that should lead to aircrew acceptability of an enhanced life-support system is ongoing.

The Air Force human systems scientists and engineers continue to fulfill the vital role of developing the technology for protecting the human operator, keeping pace with expanding operational demands. The Air Force owes much to the pioneers at Hazelhurst Field and all those who followed in their footstepts.

Aircraft/Crew Issues	Physiological Issues	Life Suport Systems
Flight Scenarios	**Hypoxia**	**Oxygen Systems**
Flight altitude	Oxygen concentration required	**MSOC Volume/Flow Requirements**
Duration of flight	Oxygen pressure/flow schedule	Concentration schedules
Cabin Pressurization	Time of exposure to altitude	Filtered air bypass
	Breathing resistance	Purge valves
Cabin pressure differential	Breathing pressure swings	Sensors/indicators
Cabin volume	**Decompression Sickness**	**LOX Converters**
Size orifice		Supply requirements
Decompression rate	Bends, chokes, CNS disturbances	Storage requirements
Transient cabin pressure after RD	Exposure times	Indicators/regulators
Flight altitude required post RD	Prebreathe requirements	Gauges/heat exchangers
Human Factors	Workload	**Gas High Pressures**
	Positive Pressure Breathing	Supply requirements
Performance degradation		Volume/flow requirements
Workspace limitations	Pressure/flow requirements	Backup to MSOC
Equipment acceptability	Cardiovascular effects	Emergency oxygen
Fit/Function of equipment	Relative gas expansion	**Regulators**
Crew comfort	Hyperventilation	Concentration schedules
Crew safety	Pressure breathing limits	Pressure schedules
	Mask vs. intrathoracic pressures	Diluter demand vs. 100 percent
	Pulmonary overpressure	Delivery rates
	Gas embolism	Inlet/outlet pressures
	Pneumothorax	Breathing resistance
	Training	Oscillatory behavior
	Ebullism	Relief valves
		Panel/seat/man mounting
	Altitude/duration of exposure	Vest/no vest press sched
	Unpressurized areas of body	Indicators/connectors
	Short-/long-term effects	Shut-off valves
	Trapped Gas	**Masks**
		Retention capabilities
	Pulmonary overpressure	Auto/manual tensioning
	Rapid decompression	Mask cavity pressures
	Delayed ear block	Breathing resistance
	G-induced atelectasis	Pressure compensation
	Sinuses	Quick disconnect warning
	GI tract	Comfort
	Thermal	**Pressure Ensembles**
		Mask/Vest/G-Suit
	Temperature	PBA Schedules
	Duration of exposure	PBG Schedules
	Protective clothing	Post-ejection schedules
		Dual anti-G suit bladders
		Sleeved vest/venous pooling
		Fully Enclosed Mask/Helmet
		Isolation valves
		Mask/vest differentials
		Max. acceptable protection
		Partial-pressure Suit/Enclosed Helmet
		Pressure schedule 140 TORR ABS
		Full-pressure Suit/Enclosed Helmet
		Pressure Schedule 180 TORR ABS

Complexity of Issues Involved in High Altitude Flight.

Exploiting the High Ground:
The U.S. Air Force and the Space Environment

Barron K. Oder

William F. Denig

Major William B. Cade III, USAF

Abstract

Since the 1930s, the U.S. Air Force and its predecessor organizations conducted state-of-the-art research into natural events in the atmosphere that affected air operations. Because atmospheric conditions frequently inhibited or prohibited Air Force missions during World War II, the service deployed its scientists and engineers to study, understand, and overcome these obstacles to national defense. After that war, Air Force experts developed many advanced solutions to atmospheric challenges, including an understanding of how solar activity interfered with high-frequency radio transmissions. At the same time, the Air Force conducted research of the upper atmosphere in anticipation of supersonic jet and guided missile operations. Before the start of the space age in October 1957, Air Force scientists and engineers had conducted research to characterize and understand the space environment. During the 1960s, space weather capabilities began to evolve into a separate expertise, largely in response to the North American Aerospace Defense Command's mission. Ever since, the Air Force, in cooperation with the Department of Commerce, has consistently added new capabilities for understanding and forecasting space weather. These accomplishments built upon one another over time to create some of the most sophisticated space weather systems that support operations by the world's finest air and space force.

An often-overlooked aspect of the success of the evolution and development of the United States Air Force's air power is the effect of weather on flight operations. Perhaps an even less appreciated aspect of the Air Force's warfighting capability has been the effect of the space environment upon Air Force systems that operate within and through space. It is now recognized that without a thorough understanding of the harsh realities of the space environment the technological superiority of the United States may be compromised. The ability of the U.S. Air Force to fully exploit the advantages *of space, often called the ultimate high ground, was the result of science and technology investments in understanding the space environment. Recent advancements in specifying and forecasting the space environment has led to a realization that space weather is the high-altitude counterpart of terrestrial weather. Yet space weather is in its infancy when compared with terrestrial weather due in large part to the complexity of the near-Earth space environment and the general paucity of measurements. A maturation of space weather requires a national strategy in which the Air Force has and is playing a key role. Like many other new areas of research, the history of space weather is a tale of cooperation between America's civilian and military scientists and engineers.*

EVOLUTION OF TERRESTRIAL WEATHER

Americans have a long history of interest in weather conditions. In what would become the United States, the earliest records of regular weather condition observations date to 1644, more than 125 years before the Declaration of Independence. Thomas Jefferson, remembered for his work on the Declaration in 1776 and as the third President of the United States from 1801 to 1809, was also considered a weather expert in his day and often responded to questions about American weather and climate[1] based on his almost unbroken series of weather observations from 1776 to 1816.[2] A lack of adequate tools hampered these early weather enthusiasts but did not diminish their interest and pursuit of knowledge and the advancement of science. Benjamin Franklin, for example, charted the Gulf Stream remarkably well from observations obtained during his many trips across the Atlantic Ocean, tried to study a whirlwind while riding a horse, and plotted the movement of a hurricane after evaluating reports from fellow postmasters.[3]

During the War of 1812, the United States government made its first entry into routine weather data collection. In 1814, Dr. James Tilton, Surgeon General of the U.S. Army, directed hospital surgeons to observe the weather and keep climatological records because of widespread interest in finding a relationship between weather and health.[4] Dr. Joseph Lovell, Dr. Tilton's successor, ordered Army surgeons at hospitals throughout the nation to prepare reports which outlined the climate, diseases most prevalent in the vicinity, their most probable causes, and the general state of the local weather-temperature, wind, rain, etc. Initially using only thermometers and weathervanes, by the early 1840s Army surgeons eventually added rain gauges, barometers, and hygrometers. In 1842, Congress appointed the first Meteorologist to the U.S. Government and assigned the position to the Surgeon General's Office.[5]

The U.S. military continued to serve as the nation's leader for weather data collection and dissemination through the Civil War and later. Acting in accordance with directions from Congress, the War Department, starting in 1870, collected meteorological observations at the military stations in the interior of the continent and at other points in the states and territories of the United States, and for giving notice on the northern lakes and on the seacoast, by magnetic telegraph and marine signals, of the approach and force of the storms. The Army responded by establishing a school of instruction in meteorology that helped train troops for their new weather-related duties. In 1870, the first bulletin announcing storms on the Great Lakes was published. In January of the following year the first 'weather probabilities' were published, offering forecasts three times daily for eight regions in the nation.[6]

On 1 October 1890, Congress directed the U.S. military to turn over its infrastructure (which included 178 weather stations, of which 26 featured automatic instruments), personnel, and responsibility for weather duty to the newly created U.S. Weather Bureau, then part of the Department of Agriculture, effective 1 July 1891. This transfer effectively ended the military's role in the nation's weather service, except for a limited capability to provide ballistic data for artillery firing, until America's entry into World War I in 1917.[7]

WEATHER EFFECTS ON SYSTEMS: WORLD WAR I THROUGH THE GREAT DEPRESSION

During the years before America's declaration of war, European belligerents quickly learned that weather conditions not only affected ground and naval operations, but also the newer weapons of modern warfare. For example, at the Battle of Jutland, the only major surface engagement between the British and German battle fleets during World War I, bad weather prevented German zeppelins from flying scouting missions prior to battle. This lack of reconnaissance kept the Germans unaware that the greatly superior British fleet had put out to sea and had every intention of surprising the German fleet in what the British hoped would be the decisive engagement of the war.[8] On 31 January 1915, the Germans unleashed the first use of poison gas during the Battle of Bolimov on the Russian front. Because of extremely low temperatures on the battlefield, the gas was so ineffective that the Russians did not report the gas attack to the other Allies. The Russian failure to communicate the news about the poison gas attack made the first use of the weapon by the Germans on the Western Front during the Second Battle of Ypres in April 1915 all the more shocking to the Allies.[9] Another weather-related limitation on the use of poison gas was more obvious: gas could only be used when the wind blew in the proper direction for a useful length of time.

After America entered the war, weather became one of the nation's most challenging foes, and bad weather often spoiled the plans of America's Airmen in 1918. In May 1918, rain caused the loss of six of the Ninety-Sixth Aero Squadron's seven aircraft when the planes had to admit they could not only not find their target, but that they were lost and had to land behind German lines. Even strong-willed air power advocate Colonel Billy Mitchell could not conquer the weather. As part of the major American St. Mihiel offensive, Mitchell had planned to conduct the largest air show of the whole war 1,481 aircraft, 609 of which were American. However, heavy rains from early September through the signing of the Armistice on 11 November 1918 meant formation flying by bombers was impossible. Under the best of conditions, mud made airfields almost useless, which meant the formations that managed to take to the air were small and unable to withstand aggressive attacks from intercepting German fighter aircraft.[10]

Because of these experiences in how weather could impair or halt the most promising operations, America's warfighters appreciated the need for further integration between tactics and meteorology. One Department of Defense (DoD) study noted that, as of the beginning of the Second World War, all the military forces of the world had more or less adequate meteorological services. At the same time, civilian meteorology had likewise gained in experience, but the problems that confronted the military meteorologists differed in many respects from those encountered by their civilian counterparts, and a great deal of research by military and civilian experts would later have to be done at the height of World War II to help secure final victory over the Axis powers.[11]

The reasons for the shortcomings in weather services were several. In the years after the First World War the United States reverted to its isolationist tendencies, endured the hardships of the Great Depression in the 1930s, and made no meaningful investments in air power or weather forecasting techniques nor equipment.[12] According to one recent study, During the depression, careful thought had to go into the development of technology before awarding a contract for aircraft purchase. Because rapid aeronautical advances and new ideas for tactical employment had created demand for an improved version, a new aircraft was often obsolete by the time it hit the field.[13] Attempts to create a separate United States Air Force met with limited success, despite Brigadier General Billy Mitchell's successful demonstration in 1921, when he and aircraft under his command sank the German battleship Ostfriesland, a prize of war from World War I. Mitchell's subsequent demotion to colonel and court-martial in 1925, for his outspoken opposition to the status quo in the military's hierarchy, likewise failed to achieve the establishment of a separate air force.[1]

When the Army Air Corps began to deliver the nation's airmail in February 1934, it was ill-prepared to do so. In less than one month of poor flying conditions, ten Air Corps pilots lost their lives,[15] scant recompense for the claim that despite the many crashes and deaths, not one pound of the 777,389 pounds of mail flown was ever lost.[16] The Air Corps took the brunt of the blame for the serious degradation in mail service, and few accepted the explanation that flying conditions in February 1934 were unusually harsh and that

General Billy Mitchell demonstrated that aircraft could sink a battleship, but the weather hindered his efforts to stage the largest offensive during World War I.

sorrowful meteorological and ground support hindered Air Corps effectiveness.[17] Based on results or on impressions formed from media reports, this additional duty failed to demonstrate to any decision makers the value of, or the need for, a separate U.S. Air Force.

Nevertheless, several important events, discoveries, and inventions marked this period. In 1924, Edward V. Appleton led an effort to verify the 1902 theory of Oliver Heaviside and Arthur Edwin Kennelly, who independently and almost simultaneously theorized the existence of the ionosphere, which allowed radio waves to bend with the earth's curvature. Appleton's 1924 work, in cooperation with the British Broadcasting Corporation, proved the existence of the E-layer of the ionosphere and measured the layer's height. During an additional experiment in 1926, Appleton discovered an upper area of the ionosphere, called the F-region. He went on to study radio-wave propagation in the ionosphere, which ultimately made significant contributions to the timely invention of radar during World War II.[18]

WEATHER, BOMBERS, AND ROCKETS–WORLD WAR II RAISES THE ALTITUDE

World War II demonstrated the versatility and power of a modern air force. From the role of aircraft in the innovative German blitzkrieg to aircraft carrier operations over vast stretches of the Pacific Ocean to the strategic bombing campaign against the Axis powers, America's Army Air Forces made significant contributions to the ultimate Allied victory in 1945. Events during the war showed that weather forecasts for military purposes relied more on past weather trends than provided a reliable, accurate forecast of what tomorrow's weather might bring.[19] Tides played an important role in deciding when amphibious operations could most likely meet with success. The most widely circulated account of the importance of weather to an Allied operation is the weather forecast for the Allied invasion of Europe at Normandy, Operation Overlord, on 6 June 1944.[20]

World War II also provided other clues to the effects of the Sun and the space environment on man-made systems. Interruptions in high-frequency radio signals hinted at ionospheric influences. British radar operators, in February 1942, encountered such severe interference that they feared the Nazis were actively jamming British radar stations as a novel type of preinvasion bombardment. However, British scientists discovered that the Sun was a powerful and highly variable radio transmitter and that sunspots and other forms of solar activity were producing potent radio emissions. The enemy of the radar was not the Germans but the Sun.[21] This was the first time scientists had proved the sun was a source of radio waves, which had many important implications for future systems.

The operational altitude of America's most advanced bombers, the B–17 Flying Fortress and the B–29 Superfortress, rose to approximately 35,000 feet and 33,600 feet, respectively.[22] At these altitudes, pilots encountered a number of weather-induced events that impaired mission success. Lightning strikes on aircraft mostly caused incidental damage, but observations by eyewitnesses in a few exceptional cases suggest that lightning caused explosions and destruction of airplanes. Atmospheric electricity, as one DoD researcher recalled, was also a concern to the ordnance experts when manufacturing powders and explosives, because of the explosion hazards connected with static discharges. The relatively high-speed aircraft used in World War II frequently had their communications and navigation aids degraded by precipitation static, a problem only partially solved in the form of discharge wicks which became standard aircraft equipment. Improvements in radios, antenna designs, and other advances during the 1940s and 1950s helped, but researchers knew that as aircraft and missile speeds go up, static charging goes up even more rapidly, and the problem of how to get rid of it equally fast remained a challenge to researchers in the early 1950s. In short, the DoD realized that even if we were never to fly anything at a higher ceiling than the 30,000 feet of World War II, the atmospheric layers of 60 to 100 miles up would be of vital importance for military operations.[23] Years later, a similar arcing problem faced spacecraft, and Air Force researchers helped lead the way in finding a solution for the nation's space vehicles.

Even with the help of early weather watchers like Thomas Jefferson and Benjamin Franklin

High-Altitude bombers such as the B–17 Flying Fortress (above) and the B–29 Superfortress (top) encountered new atmospheric conditions that often hindered operations. [Photos from U.S. Air Force Museum web page.]

and numerous others who followed them over the years until the end of World War II, no accurate method was available to forecast the weather for the next few days. More data still needed to be collected, and computers did not yet fully support the types of calculations and displays required to create meaningful weather projections. Modeling of weather events is a fairly recent breakthrough. Interest in weather forecasting by both military and civilian agencies kept researchers hard at work and still keeps them looking for newer, smarter ways of gathering, applying, and sharing their work and advancing the interests of the nation. Once man-made satellites entered the equation and started photographing wide areas of weather patterns and taking other measurements, the ability

to forecast weather improved dramatically. But even today, with fast computers and space assets, it is not yet possible to offer a warning well in advance of some weather events, such as a tornado. This type of prediction tool would have helped the nation on 1 September 1952 when a tornado made a direct hit on the 7th and 11th Bomb Wings' new B-36 bombers at Carswell AFB, near Fort Worth, Texas.[24]

If a similar pattern holds true for the ability to research and develop the means to accurately forecast space weather in a timely manner, the Air Force and its partners the National Weather Service, National Oceanic and Atmospheric Administration, and academia have made a tremendous start. The laws of science and basic research have not changed dramatically despite the advent of the space age and fast, capable, and affordable computers. It takes time to gather enough data to start modeling and simulation work of an event, and it takes more time to refine and update models and simulations of natural events. Using the technologies gained in World War II-jet aircraft and rockets-as the starting point for modern investigations into the space environment and space weather, the Air Force has been on the leading edge for almost 60 years.

Because of many aerospace breakthroughs during the World War II years, the Air Force would soon be flying higher and faster than ever before. The Second World War saw the first steps leading to jet aircraft and intercontinental ballistic missiles, both of which significantly affected the subsequent history of the U.S. Air Force and the world during the Cold War and beyond. While Germany was the only nation to field an operational jet fighter and long-range rocket weapon (the V–2) during the war, these innovations marked a new era in air power, with a concomitant demand for a greater understanding of the space environment, including space weather.[25]

These two breakthroughs-jet aircraft and missiles opened the door for a new push into understanding the natures of the atmosphere and ionosphere and their impact on current and future Air Force systems. Army Air Forces leaders appreciated the work of American scientists and researchers during World War II and took steps shortly afterward to ensure they had the people and resources to explore these new realms. The Army Air Forces, because of the teamwork of General of the Army[26] Henry H. (Hap) Arnold and Dr. Theodore von Kármán, renowned head of the Guggenheim

Aeronautical Laboratory at the California Institute of Technology, helped pave the way for the Air Force to become tightly bound to the opportunities presented by state-of-the-art science and technology. These two men stood in contrast with other significant thinkers, including Dr. Vannevar Bush, the first chairman of the National Defense Research Committee who became Director of the Office of Scientific Research and Development in 1941.[27] Bush and others like him believed that the military services should confine themselves to improving existing weapons and leave new scientific ideas to the civilian experts. General Arnold and Dr. von Kármán, however, staked out a role for military research and produced a revolutionary roadmap for the Air Force's future in terms of its participation in innovative research and development on air and space systems.[28]

While General Arnold and Dr. von Kármán staked a theoretical claim for the Army Air Forces in establishing a vital high-technology base for the future, the Army Air Forces acted to make this vision a practice. With the official surrender of the Japanese aboard the USS *Missouri* on 2 September 1945, many anticipated the quick demobilization of America's military. An even more rapid demobilization occurred among the civilians who had created the weapons the Allies used to win the war. Recognizing this potential brain drain, the Army Air Forces decided to recruit wartime personnel with expertise in electronics and geophysics for postwar employment in military research. Therefore, shortly after the disbanding of the Radiation Laboratory at the Massachusetts Institute of Technology, the Army Air Forces issued a directive to begin recruiting postwar scientists and engineers on 20 September 1945. This marked the beginning of the Cambridge Field Station in Massachusetts, which later became the Air Force Cambridge Research Laboratories (AFCRL),[29] and a new era in what would become Air Forcesponsored geophysics and space environment research, an effort that continues to this day at the Air Force Research Laboratory. Many of the personnel who formed the core of AFCRL came from the Massachusetts Institute of Technology (MIT) Radiation Laboratory, which had its origins in developing microwave radar systems essential to Allied victory in World War II. At the time of its establishment, AFCRL was located next door to MIT in Cambridge.[30]

The original headquarters of Cambridge field station, later the Air Force Cambridge Research Laboratory, located at 224 and 230 Albany Street, Cambridge, Massachusetts, as seen in 1945.

THE AIR FORCE AND POSTWORLD WAR II RESEARCH ON THE IONOSPHERE

Major General Curtis E. LeMay, the Army Air Forces' Deputy Chief of Air Staff for Research and Development, considered space operations to be an extension of air operations. LeMay challenged Project RAND, then a part of the Douglas Aircraft Company's Engineering Division in Santa Monica, California, to quickly complete a feasibility study for space operations. RAND released this study, Preliminary Design of an Experimental World-Circling Spaceship, on 2 May 1946 a mere "two days before a critical review of the subject with the Navy.[31]" The team at RAND reported, "We have undertaken a conservative and realistic engineering appraisal of the possibilities of building a spaceship which will circle the earth as a satellite." The authors noted that "such a vehicle will undoubtedly prove to be of great military value," mostly for its observation and reconnaissance potential. Even without future technological breakthroughs, such as atomic energy, the project could take up to five years and cost $150 million. The study also noted the first nation to place a satellite into Earth orbit would win tremendous respect from other nations and "inflame the imagination of mankind.[32]"

Because it would be slightly more than eleven years before the first man-made satellite did circle the Earth, it is impossible to know whether or not this vision, based on V–2 rocket technology, could have worked. But the authors of the report were prescient in at least one respect:

> *The craft which would result from such an undertaking would almost certainly*

do the job of becoming a satellite, but it would clearly be bulky, expensive, and inefficient in terms of the spaceship we shall be able to design after twenty years of intensive work in this field. In making the decision as to whether or not to undertake construction of such a craft now, it is not inappropriate to view our present situation as similar to that in airplanes prior to the flight of the Wright brothers. We can see no more clearly all the utility and implications of spaceships than the Wright brothers could see fleets of B–29's bombing Japan and air transports circling the globe.[33]

At the end of World War II, then, many separate streams converged: interest in weather conditions around the world; the future of military research and development in relation to such work in industry and academia; rapid technological breakthroughs as a result of the United States' having become the "Arsenal of Democracy"; a new willingness to understand the atmosphere; and faith that rocket technology (perhaps coupled with atomic energy) was ready to conquer the challenges of spaceflight, whenever the national will demanded it. The advent of higher-flying aircraft, observations of the jet stream in World War II, and the beginning of the rocket era had major impact. In part, the fierce rivalry between the "free world" and the Soviet Union and the role of technology as a way to gain respect and followers during the Cold War helped spur America's willingness to conquer space.

During the early years after World War II, the Army Air Forces (which became the United States Air Force in September 1947) continued its pioneering work in developing the tools necessary to explore the upper atmosphere and the ionosphere, especially in light of the new age of rockets, initiated with the appearance of the V–2. Using captured V–2 rockets and German scientists familiar with the V–2, the United States began a series of rocket firings that had both military and scientific missions and which "gave American scientists an early opportunity to develop rocket research techniques as well as make some fundamental scientific discoveries at high altitudes." The U.S. Navy had established a Rocket-Sonde Research Branch within the Naval Research Laboratory, but it had no V–2 rockets. The U.S. Army, which had 300 boxcars full of V–2 parts and equipment, had the potential to assemble dozens of V–2s, but it lacked an organized group of scientists. Therefore, in early January 1946, the Army and Navy, and joined later by the newly independent Air Force, cooperated in laying the groundwork for future V–2 rocket launches from White Sands Proving Ground (WSPG) (its current name is the White Sands Missile Range), New Mexico.[34] Russia had also captured a number of V–2 rockets and scientists. This added to the sense of importance of the work at WSPG because of the growing tensions of the Cold War. More and more people were coming to the same conclusion that the 1946 RAND report had reached: the first nation to launch a satellite from a rocket would gain the admiration and respect of the world and demonstrate a palpable superiority over its rivals.[35]

At WSPG, the American team conducted its first-ever static test firing of a V–2 rocket on 15 March 1946, with the first flight following on 16 April 1946.[36] On 22 August 1946, the newly created Air Force Cambridge Research Laboratories took part in its first V–2 launch and experiments, "Day Airglow (Photometric Technique); Propagation (Retardation and Bi-Polar Probe Methods), and Pressure, Temperature, and Density Measurements." As was true for almost half of the other V–2 missions at WSPG, officials declared this mission a failure. Because the V–2 vehicle failed, none of the experiments on the rocket could be completed. In late November 1946, the AFCRL team tried again, and with a properly operating V–2, they succeeded in getting at least partial test data from all three experiments run during the rocket's flight.[37]

The Army Air Forces launched captured German V–2 rockets from the White Sands Proving Ground, New Mexico, starting in 1946. Scientists at the Air Force Cambridge Research Laboratories flew instruments on select V–2s to probe the upper atmosphere and measure its pressure, density, and temperature.

A review of AFCRL's early V–2 flights shows a number of experiments designed to gather a variety of data like that gathered on the first rocket launch as well as solar spectroscopy, magnetic fields, composition studies, cosmic rays, micrometeors, and solar constant measurements. In total, AFCRL fired more than a dozen V–2s of the more than 60 fired by all the services at WSPG.[38] During these V–2 flights, AFCRL scientists, who had an interest in the neutral composition of the atmosphere to support Air Force missions at altitudes above 35,000 feet, measured the atmosphere's density at various altitudes. The highest altitude reached by an AFCRL V–2 was eighty-five miles on 31 August 1950. AFCRL researchers also increased the midsections of seven V–2s "by about 65 inches" to allow the addition of "scientific instruments and thus allow a larger set of simultaneous measurements to be collected for study." The data collected in these V–2 flights helped shape the scientists' understanding of the atmosphere for years to come.[39]

The growing consensus among the services' scientists held that V–2 rockets lacked the stability necessary to serve as effective test platforms. Nevertheless, by the time of the last AFCRL V–2 launch on 22 August 1952, researchers had amassed an important amount of data. A record of all three services' rocket experiments states,

> V–2s took the first solar ultraviolet spectrograms above the Earth's ozone layer. They captured spectacular photographs of Earth from high altitudes; they brought back air samples and cosmic-ray measurements. Although valuable atmospheric data were obtained, it is more

honest to regard the series of [V–2] flights as scientific test vehicles upon which new instrument and telemetering techniques were perfected. Experimenters learned how to build compact, rugged, reliable equipment, while rocket engineers found how to give the instruments a smooth, clean ride. Advances were also made in instrument pointing and recovery. This was technology rather than science,…but experience with the V–2s provided just what American scientists and engineers needed to build sounding rockets tailored specifically to space research.[40]

With the shortcomings of the V–2 more apparent, AFCRL and others interested in researching the ionosphere and beyond did indeed turn to developing sounding rockets and newer, ever more sophisticated payloads of scientific instruments to advance their work. The shift in vehicles to sounding rockets relocated much of AFCRL's work from WSPG to Holloman AFB, near Alamogordo, New Mexico. AFCRL also shifted to a new sounding rocket, the Aerobee, with its first Aerobee-borne experiment launched on 16 September 1949. However, the Aerobee sounding rockets showed some early developmental problems which, coupled with some equipment anomalies, led to many of these early experiments failing to collect useful data. Within a year or so, though, AFCRL began compiling a sterling record of successful experiments. Between 1952 and 1958, AFCRL's experiments used a series of different sounding rockets-improved Aerobees, such as the Air Force Hi-Aerobee featuring a "nose cone containing data-recording instruments" capable of being "recovered by parachute,"[41] and

the Nike-Deacon and Nike-Cajun rockets.[42] These sounding rocket experiments collected valuable data and also paid other dividends. By the late 1950s "Thirty colleges, universities, and research institutes were participants in the Air Force study program on the relationship between weather and solar activity.[43]

These and other sounding rockets allowed affordable access to space for experiments

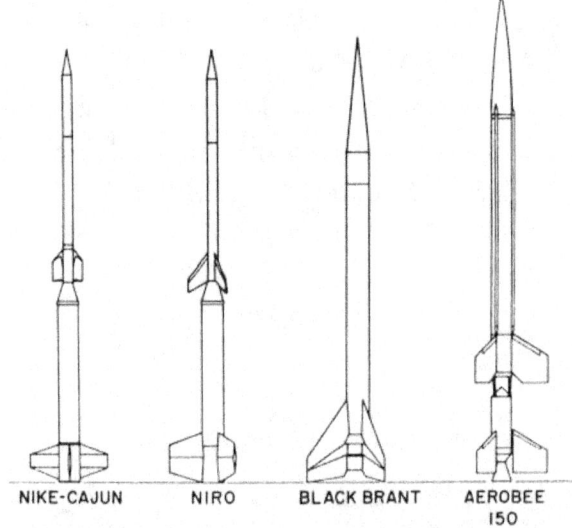

Line drawings of a few of the types of sounding rockets Air Force experts used to study the upper atmosphere and some of the ground-based trackers used in conjunction with various sounding rockets.

and near-Earth observations. Because these were suborbital rockets, experiment packages routinely did not achieve Earth orbit and their payloads were relatively small, though the Nike-Deacon was a two-stage rocket. AFCRL experiments helped researchers understand the operational environment for the latest generation of high-altitude jet aircraft as well as the new field of missile guidance applicable to the latest weapon in the nation's arsenal, the Intercontinental Ballistic Missile (ICBM). In just ten years, Air Force and other researchers working with sounding rockets had raised "the engineering standard atmosphere…from 20 to above 100 km [kilometers]," developed a means of photographing weather over wide areas that "applied to the study of hurricanes," and developed newer, more accurate telemetry packages as part of these early efforts. In addition, sufficiently accurate measurements have been made of the altitude variation of pressure, cosmic-ray intensity, solar X-ray intensity, and airglow which could "be used as the basis for moderately accurate absolute altimeter systems."[44] AFCRL researchers, through their V–2 and sounding rocket work, helped the nation amass the data that brought about "the basic knowledge of the upper atmosphere and the technology necessary for the later planetary exploration and space science, as well as for the military development of ICBMs and supersonic aircraft."[45]

THE AIR FORCE, THE VAN ALLEN BELTS, AND THE START OF THE SPACE AGE

Arguably, though, the most significant consequence of the sounding rocket work by AFCRL and other pioneers became clear only after years of work. "The ability to launch an artificial earth satellite by means of the Vanguard vehicles is a direct outgrowth of experience gained in performing upper-air rocket experiments."[46] Before the Vanguard vehicle successfully lifted its first satellite into orbit on 17 March 1958, the Soviet Union had launched both Sputnik I on 4 October 1957 and Sputnik II on 3 November 1957, and the United States had launched its first satellite, Explorer I, on a Jupiter–C launch vehicle on 31 January 1958, a direct descendant of the German A–4 (V–2) rocket. The third U.S. satellite launch, Explorer III on 26 March 1958, confirmed the existence of the Van Allen belts.[47] These initial successful launches of man-made satellites represented varying degrees of significance for the International Geophysical Year (IGY). The IGY ran from July 1957 through December 1958 and "was timed to coincide with the high point of the eleven-year cycle of sunspot activity."[48] A remarkable demonstration of the willingness to share information during the Cold War, the IGY ultimately involved researchers from sixty-seven nations for "cooperative study of the solar-terrestrial environment.[49]

The first two Soviet satellites were relatively large and did not return much useful scientific data. Nonetheless, the mere fact that these were on orbit earlier than any satellite from the "free world" gave prestige to the Soviet Union, exactly as the 1946 RAND study on the world-circling satellite had predicted. Ironically, though, the smallest of the first three man-made satellites, Explorer I, contributed the most to the IGY and future of space systems in its discovery of what were later called the Van Allen belts.

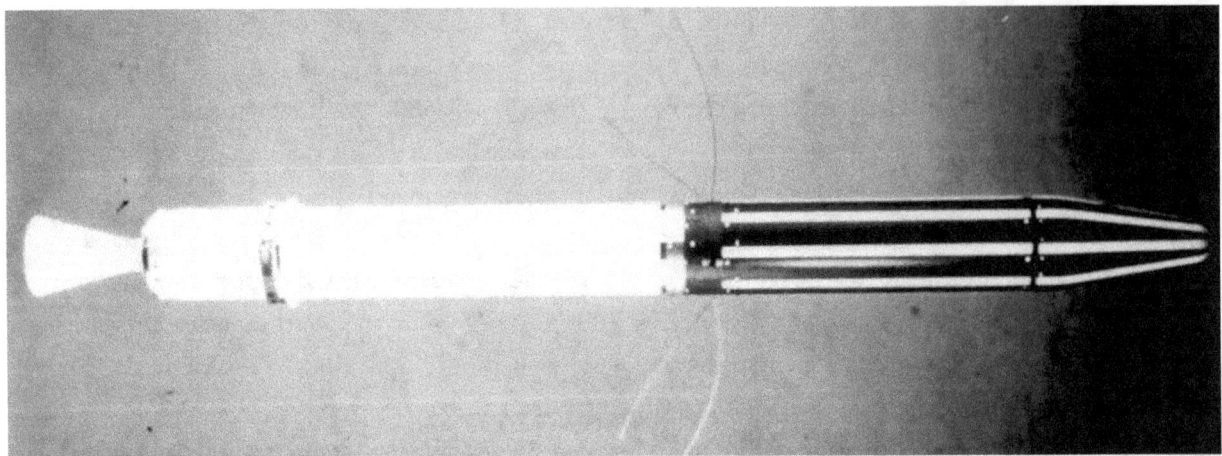

Explorer I, officially known as Satellite 1958 Alpha, was the first U.S. satellite, and third satellite overall, to obtain Earth orbit. Explorer I carried the first AFCRL, experiment into space, making the start of the Air Force's prominent role in space. Once in orbit, Explorer I indicated a much lower cosmic ray count than Dr. Van Allen had expected. He theorized the existence of a belt of charged particles trapped in space by the earth's magnetic field. The existence of these belts, later named the Van Allen belts, was confirmed by measurements taken during the flight of Explorer III.

Explorer I established "a lasting American superiority in miniaturized electronics.[50]"

On 1 May 1958, Dr. James A. Van Allen of the State University of Iowa and his team of researchers announced that data taken from scientific equipments on Explorer I and Explorer III "disclosed an unexpected band of high-intensity radiation extending from 600 miles above earth to possibly an 8,000-mile altitude." Dr. Van Allen said the radiation was 1,000 times as intense as could be attributed to cosmic rays.[51] The discovery of the inner Van Allen radiation belt opened a new area for Air Force research: the magnetosphere. This is the part of space that picks up where the ionosphere leaves off and "is filled with magnetic fields, electric fields, matter, energy, and activity invisible to the naked eye but readily apparent to more sensitive scientific instruments.[52]"

People have used the magnetic compass since its discovery in China in approximately 1000 AD. In 1600, William Gilbert concluded the compass worked because "the Earth is a giant magnet.[53]" The Earth, resembling a dipole bar magnet, exerts a magnetic force field that controls the magnetosphere. The magnetosphere, encompassing the inner and outer Van Allen belts, is where manned and most unmanned spacecraft operate. The many space vehicles that have probed this region have more clearly mapped its shape and elucidated its varying conditions and have laid the groundwork for exploiting modern space-based systems.[54]

The discovery of the Van Allen belts and the exploration and mapping of the magnetosphere for more than a decade point to an integral aspect of scientific research and technology development: time is the first ingredient. For years, scientists had theorized the existence of a protective barrier around the Earth shielding us from the many solar irregularities observed since the time of Aristotle. Ground-based observations had yielded a good start, and the V–2 and sounding rockets added to this collective knowledge from about 100 to 130 miles in space. Sounding rockets provided Dr. Van Allen and his colleagues with a solid understanding that the atmosphere thinned with increasing distance from Earth's surface. Logically, then, he expected that the amount of cosmic radiation, fatal to humans if unattenuated atmospherically, would increase as one's altitude in space increased. Therefore, when Van Allen sent his first Geiger counter into orbit on Explorer I, he had already surmised the existence of a radiation belt surrounding the globe.[55]

When Van Allen received data from the experiment, results contradicted common sense: data sent to ground stations showed a lack of radiation. With a new instrument on board Explorer III a few months later, Van Allen and others reducing the data realized the unexpected results from his first mission could be explained consistently with previous observations. At altitudes lower than 400 miles above Earth, radiation could be counted with relatively routine devices. Beyond that distance, though, "there was so much radiation…that it could not be accounted for by simple cosmic-ray bombardment." Plus, "careful analysis of the data proved that there were two roughly doughnut-shaped belts, or a single gigantic one, of trapped radiation circling Earth." This marked the first time that scientists had "defined the barrier that protects it [Earth] from the deadly barrage of celestial radiation." By the time the Van Allen belts were fully measured and characterized, scientists realized that Van Allen's initial height of the belts, 8,000 miles, was hindered by young technology. As technology matured, scientists mapped the Van Allen radiation belts out to 40,000 miles around the Earth.[56] This underscored the "push-me pull-you" nature of technology and space exploration. As technology improves, scientists are able to launch ever more-sophisticated equipment to take increasingly precise measurements to produce a better, more accurate understanding of the space environment. Because new technology is often more susceptible to the impact of the space environment and space weather, new research areas are revealed to understand and counter any negative impacts on new systems.

The discovery of the Van Allen belts opened a new door to space research. Catalyzing space exploration, in addition to the interest sparked by the IGY and the creation of the National Aeronautics and Space Administration (NASA) and the concept of solar wind articulated in 1958, was, of course, the Cold War. President Eisenhower, fearing a surprise attack from Russia or another source, had sought high-altitude aircraft for surreptitious reconnaissance flights. However, the concept of national airspace hindered spy flights over hostile territory. Both the United States and the Soviet Union had made it clear that spy planes could easily be the reason for the next war, which all parties knew could easily escalate to a full thermonuclear event.[57]

In one sense, the fact that the Soviet Union orbited the first man-made satellite offered the United States an opportunity to answer a question that had long burned within Washington, D.C.: How would a nation respond to a hostile nation's flying a satellite over the homeland? Because the United States did not declare Sputnik's path over the United States an intrusion of national airspace or a hostile act,

129

some credit the peaceful uses of outer space to President Eisenhower's measured response to the Soviet space challenge. Thus the corollary, the Soviet Union could not challenge the flights of U.S. communications and surveillance satellites over the Soviet Union.[58] This thoughtful American response also set the stage for the duality common in America's science and technology history   continue moving forward simultaneously with clearly defined military missions and civilian goals, with a minimum amount of friction and a maximum benefit to the nation. This had been true of the earliest weather forecasting efforts and continues today with respect to space environment research.[59]

INNOVATIVE AIR FORCE GROUNDBASED AND BALLOONBORNE PROGRAMS

The 1950s also marked a period of diversification in the Air Force's approach to learning about the space environment. It had been clearly understood that the Sun was the ultimate source of energy for the upper atmosphere and ionosphere. However, the causal relationship between solar disturbances and variations in the near-Earth space environment, or what eventually became known as space weather, was not well known. AFCRL experts now branched out to establish a new, state-of-the-art coronagraphic observatory at Sacramento Peak near in the felicitously named Sunspot, New Mexico.[60] In 1952 this site began basic research on "solar physics and solar-terrestrial effects." Later in the 1950s, programs in "solar radio astronomy and trans-ionopsheric propagation" began at the Peak. Later work at the site included "long-term

measurements of solar extreme ultraviolet radiation." The site later became part of the National Solar Observatory, with on-site Air Force scientists and engineers contributing to its research on the sun and its influence on modern electric equipment, including communications and navigation systems.[61]

Other work with solar-observing ground stations advanced by the Air Force over the years includes participation in the Solar Electro-Optical Network (SEON), which consists of the Solar Optical Observing Network and the Radio Solar Telescope Network. The global network of SEON sensors continuously monitors the sun at visible wavelengths and at radio frequencies to detect

The Sacrament Peak Observatory includes the Solar Vacuum Telescope (top), which features a 220-foot vertical shaft and a connecting laboratory building. The site also includes a 16" coronagraph and a 12" coelosts telescope for solar observation.

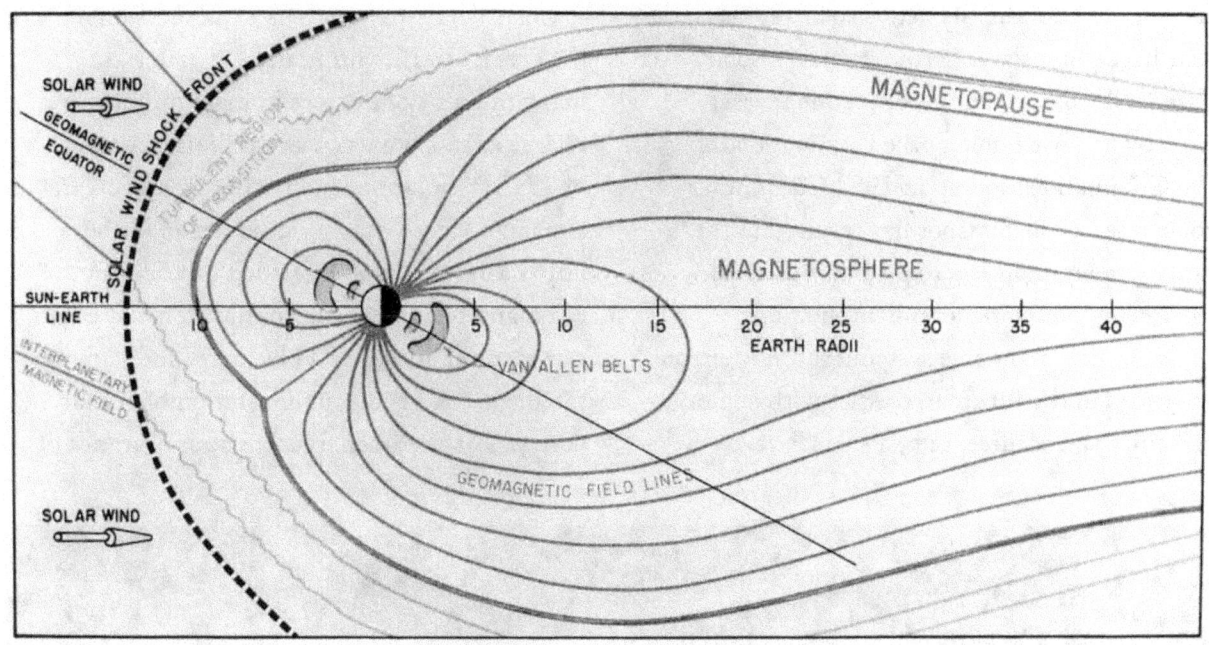

Charts created by Air Force scientists show some of their discoveries over time and their conclusions about nature and effects of the space environment.

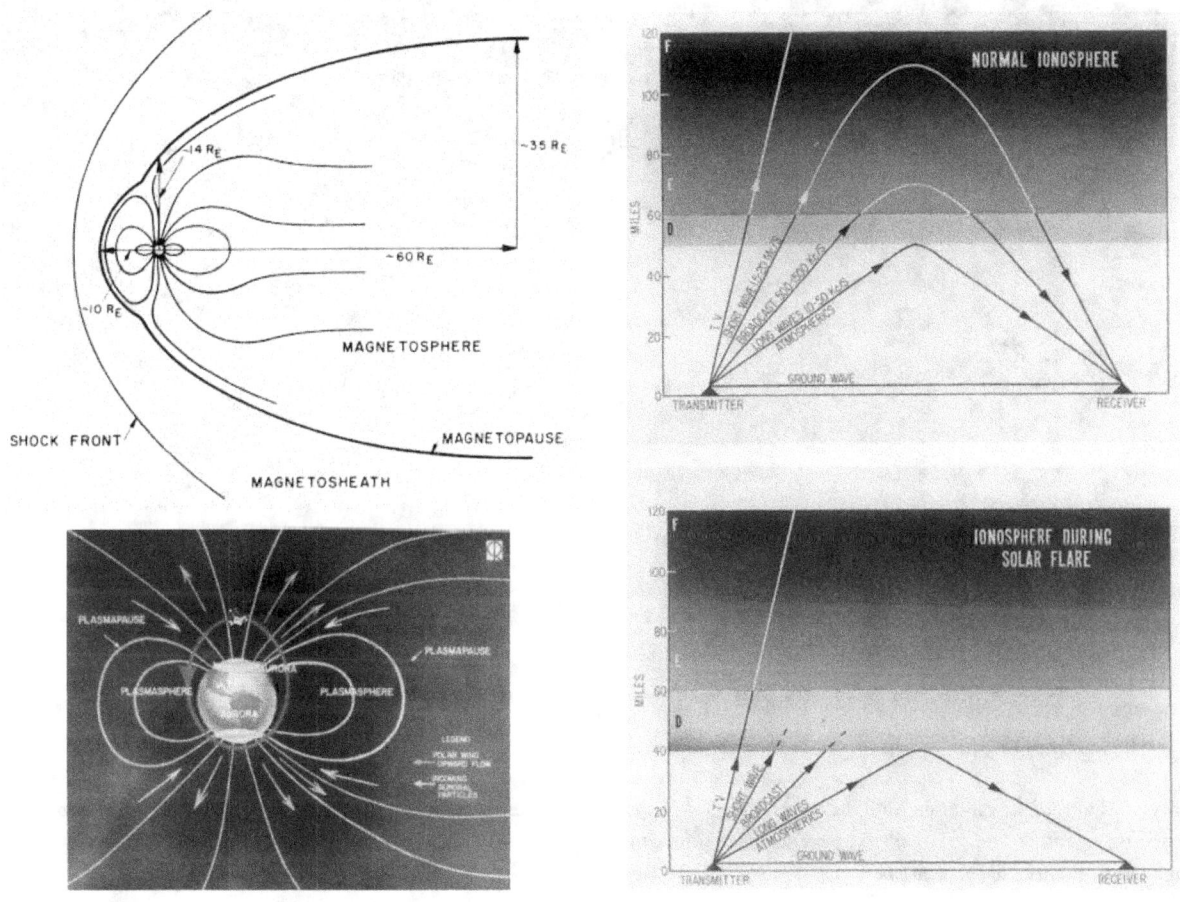

solar flares and identify active solar regions that have potential for erupting into a solar flare.[62] SEON became an operational asset used by the Air Force Space Command after the Command stood up in 1982 Continuing solar research at Sunspot has resulted in the development of an Improved Solar Optical Observing Network, a semiautonomous, remotely commandable system that may enable scientists in the future to predict with some certainty the occurrence of a solar flare.

Solar flares, tremendous explosions on the surface of the Sun, can heat material to many millions of degrees in mere minutes and release as much energy as a billion megatons of TNT.[63] Solar flares, in other words, create a massive electromagnetic explosion in the upper solar atmosphere which can disrupt radio and telephone communications on Earth, cause power surges and blackouts and damage satellites.[64] A visible manifestation of a solar flare can be the uncharacteristic appearance of

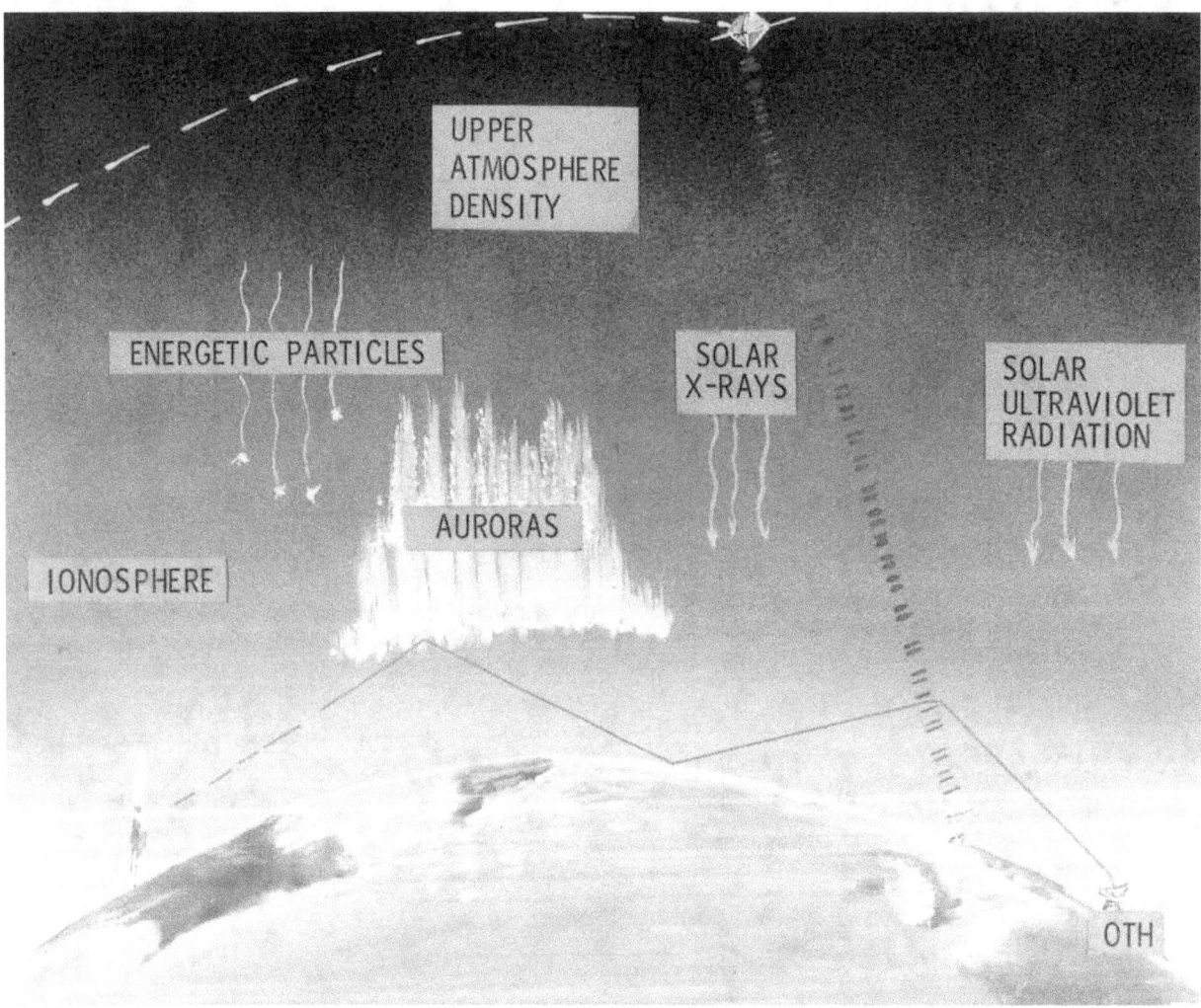

X-rays and ultraviolet radiation from the sun is absorbed in the upper atmosphere, heating it and causing it to expand. If a solar flare occurs, greatly increased numbers of energetic particles cause auroras to become much more active, changing the radio signal propagation qualities of the ionosphere.

the Aurora Borealis as far down as Illinois and Oregon, a result of solar wind and energetic particles created by a solar flare.⁶⁵ The SEON complements the work of the National Oceanic and Atmospheric Administration which "sponsors the Geostationary Orbiting Environmental Satellite (GOES) series that monitors solar flare X-ray emissions.⁶⁶" This type of cooperation between military and civilian research communities had been important in the early days of terrestrial weather observation and the evolution of weather forecasting, and it continues today in the ongoing efforts to create the ability to forecast the space weather.

These newer systems built upon pioneering work on solar flares conducted by AFCRL researchers in the early 1960s, making the laboratory "one of the earliest players in the development of 'space weather' forecasting in the US. AFCRL studied ways to identify specific 'precursors' of solar flares that, in turn, could provide some ability to forecast their occurrence." Yet other work at AFCRL "looked for the 'signatures' of the solar flares that released high-energy protons" in the hope that researchers could find a way "to provide some advance warning of disturbances that would occur at Earth within minutes or a few hours" of the observed event. This type of work led to the development of the Proton Prediction Model, which incorporated computer algorithms "to predict the occurrence and arrival time of solar energetic particles at the Earth." This innovative system "became an operational space weather tool at the Air Force Global Weather Central at the end of 1987.⁶⁷"

Another approach championed by AFCRL to understand, characterize, and map the

space environment involved the use of high-altitude balloons. In fact, Project Mogul, "the first major geophysics program that the Army Air Forces started early in 1946," looked at "the feasibility of using balloon-borne acoustic sensors for long-range detection of potential Soviet missile launches and atomic tests" during the early years of the Cold War. In 1951, when the DoD transferred the duty of creating experimental equipment for meteorological purposes to the Air Force, specifically to one of AFCRL's directorates, more balloon-related programs came to the laboratory. In the 1960s, with the nation swept up by the space race and the race to the moon, AFCRL engineers used their "expertise in balloon design and launching" created "during the 1950s with programs like Project Moby Dick, which measured upper atmospheric winds" to support the emerging U.S. role in space. Then, in the 1960s, with the nation's eyes on landing probes and ultimately men on the moon, these "balloon engineers began a new area of work, providing drop-tests for re-entry systems to be used in lunar and planetary probes.⁶⁸"

Drop-tests by AFCRL personnel supported several NASA missions starting in the early 1960s and continued until the last drop-tests in 1982. The 70 drop-tests conducted in support of the Surveyor mission to the moon required AFCRL engineers to simulate "the descent of Surveyor's lunar landing vehicle…by releasing the vehicle suspended under its open parachutes from a balloon…tethered at 1,500 feet." Conducted at Holloman AFB, these drop tests helped to verify the Surveyor spacecraft's ability to retrofire its rockets and achieve a soft landing. In July and August of 1965, similar tests on a much larger scale dropping a one-ton reentry system from 130,000 feet supported NASA's Voyager mission to Mars.[69]

With the aftermath of the first space launches in 1957 to 1958 and the frantic pace of Cold War posturing for national superiority via the number (versus the quality) of space launches, space became an ideological shuttlecock between the "free world" and the Soviet system. This clash of apparently irrevocable ideologies put space scientists in a difficult position because of America's desire to use technology as a lever for prying neutral nations away from sympathies for the Communist/Socialist system and to bring them into the fold of free enterprise and democracy. Ironically, though, America's ideological adversaries were trying to do exactly the opposite by claiming their own superior technology. Consequently, the space race afforded the U.S. Air Force myriad opportunities for state-of-the-art space research programs, a challenge the service accepted without hesitation but within the limits of peaceful exploration. These early years also established the dual paths of military and civilian approaches to the role that space would play in providing for the nation's security while simultaneously promoting scientific breakthroughs for future system designs. The Air Force and other defense-related agencies would conduct their reconnaissance missions in a relative blackout mode, while NASA pursued free and open relations with other nations providing they restricted themselves to space science and released their data to all the world. This is how the United States first built its reputation as a fair and dependable provider of launch services for other nations.[70]

THE AIR FORCE ORGANIZES FOR SPACE AND MOVES AHEAD

up to take the initiative in this transitional period in American space policy and history. This is not to say, though, that the road to the Air Force's ultimately assuming responsibility for the military's mission in space was smooth and straight. All three services made their own thrust for primacy in the space mission, but shortly after President John F. Kennedy entered office in January 1961 the Air Force reorganized and created the Air Force Systems Command and the Air Force Logistics Command. The former received responsibility for all research, development and acquisition of aerospace and missile systems, to include strategic missiles. The latter organization was established to handle maintenance and supply only. Without this type of overhaul, the Air Force could not have risen to the DoD's challenge for the service to take over future military space development responsibilities. Even with this new charter for leading the DoD's space efforts, the Air Force needed time to properly define its relationship with the other services and NASA and to reassure the public about the potential for a military presence, manned or unmanned, in space.[71]

One of the ironies of the early space age was the perception of the need for a military role in space, but often a stumbling block to achieving this was the military itself. Because the nation's strategic defense relied upon nuclear weapons (delivered from bombers, intercontinental ballistic missiles, and submarine-launched nuclear missiles) and the ability of the weapons to deter hostile nuclear action by other nations,

Over the years, Air Force reasearchers used a variety of balloons and aerostats from many different locations—including a ship from the United States Navy—and under various environmental conditions to collect data to support their state-of-the-art explorations of the atmosphere and ionosphere.

atmospheric nuclear tests were not uncommon. Researchers understood these tests could have some impact on the few satellites on orbit or upon early astronauts circling the Earth, and due precautions were taken, but high-altitude tests continued. Two events in 1962 demonstrated the dramatic changes within the radiation belts that these tests could produce. On 8 July 1962, the United States conducted a high-altitude nuclear test, and in October of that year, the Air Force announced special instruments on unidentified military test satellites had confirmed the danger that astronaut Walter M. Shirra, Jr., could have been killed if his…space flight [on 28 September] had taken him above a 400-mile altitude. The July blast caused an artificial radiation belt with peak intensities at least 100 times greater than normal.[72] Two days after this same nuclear test, the world's first communications satellite, Telstar, was launched. For six months, Telstar made history, speeding communications and making the world a smaller place by adding the phrase 'live via satellite' to the common vernacular.[73] Telstar, though, was knocked out after six months by radiation from the

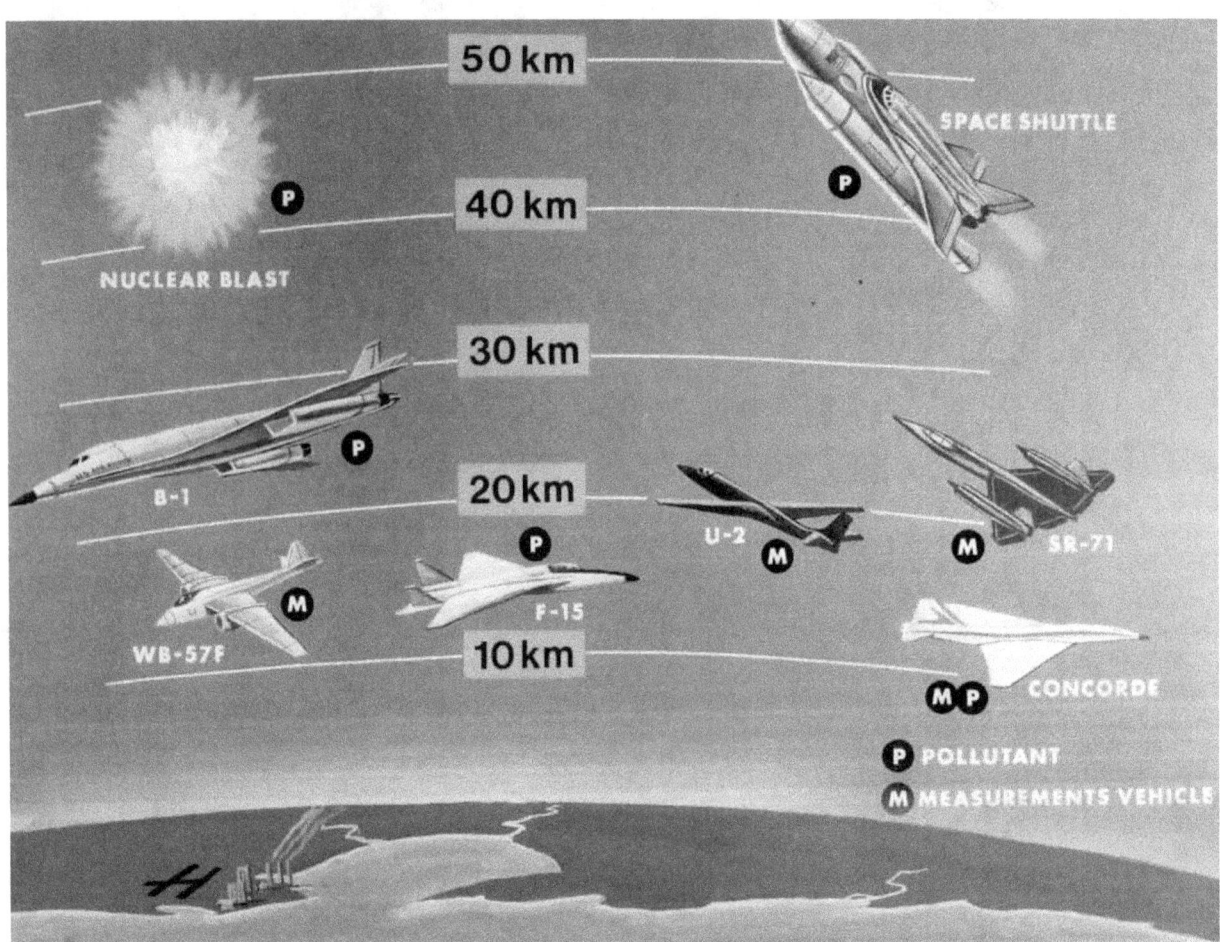

Because Air Force systems operate at different altitudes, researchers have long looked for new and better ways to characterize the various operational environments of interest to ensure maximum performance of current and future Air Force and civilian systems.

American high-altitude nuclear tests.[74] The 1963 Nuclear Test Ban Treaty banned future nuclear tests in the atmosphere and space, effectively eliminating this hazard to space operations for the United States and the rest of the world.

Nevertheless, the Air Force continued its pioneering work in understanding the rigors of space. Air Force organizations such as the Air Force Weapons Laboratory (AFWL), located at Kirtland AFB, New Mexico, played a role in the evolution of the Air Force's mission in space and researching the space environment. From its creation in 1963 through 1973, AFWL made major scientific and technical contributions to understanding natural and artificially created space radiations through a series of satellite flights and space probes. Much of this work focused on mapping the inner Van Allen belt more completely, measuring radiation levels and geomagnetic fields in both the inner and outer Van Allen Belts, and to compare measurements with theoretically predicted environments. Two other satellites carried AFWL instruments into space to study solar flares, cosmic radiation and the effects of solar storms on the Van Allen Belts. One of these satellites provided high precision measurements of polar cosmic radiation and recorded unique data on a major solar flare during May 1967 and had a significant role in defining the hazards for manned space missions during solar active periods. Later AFWL efforts helped to establish the limits on the length of time astronauts could operate in near-earth orbit behind various types of spacecraft shielding and collect data of considerable interest to both manned and unmanned operations in low- to mid-earth orbits.[75] Over the years, AFWL

The Air Force Weapons Laboratory (AFWL) put its Fiber Optics Experiment aboard the Long Duration Exposure Facility (LDEF) satellite to study long-term effects of space background radiation. After four and a half years in space, the Space Shuttle retrieved the LDEF in January 1990, with AFWL's experiment seen on the lower right hand corner of LDEF.

and its successors, like the Phillips Laboratory and the Air Force Research Laboratory, used AFWL's expertise in measuring radiation and developed it into world-class methods for radiation-hardening techniques that have helped to create and preserve America's reputation as the leader in space operations.

The most sustained contributions by an Air Force organization dedicated solely to geophysics research and development came from the AFCRL, which became the Air Force Geophysics Laboratory in January 1976.[76] The nascent AFCRL began recruiting scientists in

137

September 1945, shortly after the Japanese surrender ended World War II. Many of the programs it initiated in the late 1940s made consistent contributions to state-of-the-art knowledge about space for years. For example, AFCRL pursued programs in theoretical and applied research in meteorology, upper atmosphere, and solar studies, and ionospheric and seismo-acoustic propagation. As science advanced, in the late 1950s AFCRL added programs in space physics and optical/infrared studies, followed shortly by geodesy and gravity. An engineering program that developed experimental meteorological equipment, sounding rockets and balloons, and instruments for satellites provided AFCRL the synergy and critical mass needed for successfully researching space weather and the space environment.[77]

The discoveries made by Air Force and other researchers have constantly grown and changed our understanding of what to expect as we explore and exploit space. A look at the state-of-the-art technologies and breadth of scientific knowledge and data show both constantly expanding in absolute terms. Van Allen's initial belief in 1958 that the belts extended to 8,000 miles above Earth was just the start. Because of the work of AFWL and AFCRL, among others, scientists now know that the starting point for the inner belt ranges, depending upon latitude, from 250 to 750 miles, rises to approximately 6,200 miles, and traps protons. The outer belt, though, traps electrons and extends from the top of the inner belt...up to 37,000 to 52,000 miles depending on solar activity.[78]

THE AIR FORCE EXPLOITS THE SPACE ENVIRONMENT AND SPACE WEATHER

The space environment and space weather, taken together, represent the major challenge facing Air Force researchers in their quest to field the best systems in support of the United States. According to one recent work,

Space weather starts inside the Sun and ends in the circuits of man-made technologies. Defined simply, space weather is a range of disturbances that are born on the Sun, rush across interplanetary space into Earth's neighborhood, and disturb the environment around our planet and the various technologies-cell phones, satellites, electric power grids, radios-operating in that environment. The key to space weather is the transformation of energy, a transformation from magnetic energy and intense heat on the Sun to plasma energy in interplanetary space to magnetic and electrical energy around the earth.[79]

Just as the study of weather on Earth often levied different methods on civilian and military meteorologists, Military and civilian space weather requirements are similar but often addressed independently. In addition, the lack of fundamental knowledge about the physics of space weather and its impact on technology makes it difficult to develop and evaluate techniques to compensate for space weather. This is something Air Force and other researchers are trying to conquer through extensive basic and advanced research into the space environment in the hope of ultimately being able to forecast space weather.

Space weather is difficult to understand because it isn't tangible, like a thunderstorm or tornado.[80] Rather, space weather and the space environment are dominated by a tenuous gas made up of charged atomic particles, including protons, electrons, and ions. Another factor is the solar wind, a very tenuous plasma, having a density of only a few protons or electrons per cubic centimeter. A paradox with the solar wind is that it could not be felt if it were to blow against your face, while its velocity greatly exceeds the velocity of any winds on Earth, being over a million miles per hour with an extremely high temperature in excess of 100,000 degrees.[81]

Much of this knowledge about the space environment and space weather came from years of research and development by Air Force scientists and others receiving Air Force support, and ranged from advanced ground-based equipment to balloons to state-of-the-art space experiments and systems. After all of the work accomplished by the Battlespace Environment Division, as the once-AFCRL is now called, there remains the challenge of developing state-of-the-art systems to collect data that supports creation of quality modeling and simulation tools for the warfighter.

One such system is the Digital Ionospheric Sounding System, which brought together the expertise of the Air Force Air Weather Service (now the Air Force Weather Agency), the Air Force Space Forecast Center (now the Space Weather Operations Center), and the Geophysics Directorate of the Air Force Phillips Laboratory (now the Air Force Research Laboratory). The Research Laboratory developed this ionosonde, which sends out radio waves to sense the overhead ionosphere and measure the electron density profile. The Air Force Weather Agency currently operates the Digital Ionospheric Sounding System, which features 20 locations around the world, to gather global ionospheric data in real time. The Space Weather Operations Center ingests this collected data to create ionospheric specifications and forecasts for communications, surveillance and navigation systems.[82]

Another ground-based system, the Ionospheric Measuring System, also developed by the Air Force Research Laboratory, measures the total electron content within a sampled volume. This system uses sophisticated coordination between a ground station and a Global Positioning Satellite (GPS) to measure integrated electron density content. By taking enough of these snapshot readings over a large area, researchers hope to be able to use the readings to model the local ionosphere. These snapshots play a role in devising new atmospheric models which can evolve over time as new readings are obtained or as new technology appears to more accurately measure the ionosphere.[83] Ultimately, the hope is to create an accurate model of the ionosphere to support creating a space-weather forecasting system, just as extensive efforts to gather terrestrial weather data helped create today's weather forecasting modeling and simulation tools.

The Ionospheric Measuring System is also looking at the well-known problem of ionospheric scintillation, a phenomenon quite similar to atmospheric scintillation. Variations in the ionosphere's density are known to cause variations in signals transmitted from on-orbit satellites, such as the GPS system. Scintillation can cause satellites and ground stations to lose lock, losing messages. While current models are able to handle the effect of a GPS beam traveling through the ionosphere, extreme solar activity and scintillations that can contribute to signal degradation are not modeled.[84]

The Air Force has also funded computer-based models in academia, such as the Magnetospheric Specification Model (MSM) at Rice University in the late 1980s. It took nearly ten years for a team of scientists and graduate students to design, build, and test the MSM, which computes the intensity of energetic electrons and ions in the equatorial plane of the earth. MSM can take real-time data from satellites and ground stations, or archived data from past storms, to create a continuous model output that gives the conditions of a storm as it is actually going on. Such nowcasting is useful because it provides a picture of where the most intense features...are at any time, and of how the storm is changing.[85]

While nowcasting has a role to play in studying space weather events as they occur, Air Force and other researchers are greatly interested in developing an ability to forecast space weather and its potentially harmful effects on man-made systems. To use MSM for forecasting, researchers have incorporated artificial neural networks. These networks are an attempt to use the human brain's ability to recall and associate past sights and sounds with current and, to a limited extent, future conditions. In computer terms, Inputs to mathematical neurons within a computer can be trained to recognize and associate combinations of inputs with certain outputs. This type of new work has many caveats, but the Magnetospheric Specification

and Forecasting Model can already take this approach and provide good one-hour forecasts of Magnetospheric conditions. It can also generate longer forecasts, but the accuracy suffers.[86]

A good example of how the Air Force works with industry to create new space-weather forecasting tools is the Parameterized Real-Time Ionospheric Specification Model (PRISM). The Battlespace Environment Division of the Air Force Research Laboratory has funded Computational Physics, Incorporated of Norwood, Massachusetts, to develop PRISM, the world standard model for High-Frequency communication and satellite surveillance. The 50th Space Wing at Schriever AFB, Colorado, used PRISM to provide timely reports on global ionospheric parameters to all DoD customers in the early 1990s, while Computational Physics indicates that more recently the 55th Space Weather Squadron (part of the Air Force Weather Agency collocated at Schriever AFB) also uses PRISM in support of military communications. The firm notes that a new version [of PRISM] incorporating the plasmasphere is currently under development.[87] If you were to go to places like the Air Force Weather Agency, Dr. William Denig of the Battlespace Environment Division said, their ionospheric model, the one they use operationally, is PRISM,[88] making PRISM an important Air Force contribution to the warfighter.

The Air Force has been developing accurate and timely space-weather simulation models since the early 1960s. Recently, though, as data has become both more plentiful and detailed, the service has undertaken a new look at how to model the space environment in support of developing the ability to forecast space weather. Changes in space weather are attributable to the highly variable, outward flow of hot ionized gas (a weakly magnetized 'plasma' at a temperature of about 100,000 degrees Kelvin, called the solar wind) from the Sun's upper atmosphere. In addition, nonthermal, electromagnetic waves in the X-ray and radio portions of the spectrum also play a role in the conditions of space weather.[89]

The Geophysics Directorate of the Air Force Phillips Laboratory realized that, for the most part, space weather forecasters have been essentially blind to impending interplanetary disturbances from the time they leave the Sun until the time they impact the magnetosphere. To fill this void, in late 1994 the Directorate's Solar Wind Interplanetary Measurements system, launched on NASA's Wind spacecraft, went on-orbit and started to provide users at the 50th Weather Squadron, at NASA, and at the National Oceanic and Atmospheric Administration with advanced warning of about 1 hour related to the effects of geomagnetic storms. The Air Force Space and Missile Systems Center's Space Test Program sponsored this innovative approach to forecasting space weather.[90]

Because the solar wind carries energetic particles from the Sun, it could also cause damage to spacecraft as a result of electrical charging, much like the atmospheric electricity noted by aircraft pilots during World War II. Then, the sparks could cause communications problems and arc out of the aircraft through protuberances like engines. On modern spacecraft, however, with computers and other sensitive equipment onboard, a charge could

The Air Force Geophysics Laboratory developed the primary experiments for the Spacecraft Charging at High Altitude (SCATHA) satellite launched in 1979. SCATHA demonstrated successfully for the first time a capability to discharge spacecraft at geosynchronous altitude by means of a low-energy plasma source.

cause a loss of communications or prevent the satellite from functioning, becoming little more than an orbiting brick.[91]

Although space experts had known about spacecraft charging for some time, it became important to master this phenomenon as spacecraft began to include more sophisticated computers as their brains and electronics for performing their missions. In January 1979, the Geophysics Laboratory's Spacecraft Charging at High Altitudes (SCATHA) satellite lifted off into space. Air Force Systems Command's Space Division, through its DoD Space Test Program, managed the SCATHA satellite, with its thirteen experiments from the Air Force, Navy, NASA, Defense Nuclear Agency, industry, and university groups. SCATHA supplied data to support the development of a comprehensive specification of the space environment at geosynchronous altitudes. SCATHA, a highly successful program, succeeded in its mission to obtain environmental and engineering data to allow the creation of design criteria, materials, techniques, tests and analytical methods to control charging of spacecraft surfaces and to collect scientific data about plasma wave interactions, substorms, and the energetic ring. In 1982, Geophysics Laboratory personnel closed out the SCATHA program when they delivered to Space Division the SCATHA Data Atlas, a statistical compilation of the environmental conditions experienced by SCATHA at geosynchronous earth orbit. They also delivered a computer code that was a charging design tool for calculating potentials on and about a three-dimensional satellite composed of various materials. Finally, researchers also provided a report on the active control of spacecraft potential by the emission of plasma.[92]

Moore's Law, expounded in 1965 (three years before the creation of Intel), essentially says there will be a revolution in computer capabilities every eighteen months because engineers would be able to basically double the number of electronic devices onto microchips.[93] The law seemed to be holding true throughout the 1970s and 1980s, so the sensitivity of electronics to the space environment and space weather continued to grow along with this sophistication. By 1990, the Air Force needed to evaluate the effects of the space environment on the latest generation of electronics equipment.

The Air Force and NASA cooperated on the Combined Release/Radiation Effects Satellite (CRRES) that launched in July 1990. CRRES

brought together the Air Force Space Radiation Effects program's effort to seek better ways to shield microelectronics and NASA's Chemical Release Program, which explored using chemical releases in space to explain the near-Earth space structure. During 1990 and 1991, CRRES collected data that revised the standard NASA…models of the radiation belts that are used by satellite designers. At the same time, CRRES also had the great fortune to record the creation of a long-lasting, third radiation belt (a second inner belt of protons) around the Earth following the major geomagnetic storm of March 1991. Scientists had long debated the possibility of this third belt but had never proved its existence before CRRES.[94]

Air Force researchers developed innovative solutions to the problem of spacecraft charging in support of systems like the GPS that operate in a high-altitude or geosynchronous orbit. The charge control system launched in July 1995 and demonstrated that an autonomous, active system designed to prevent surface-charge buildup on deep-space satellites offered a way to avoid this potentially mission-ending problem. The charge control system featured a suite of charging sensors that could detect the start of, and source of, surface charging. A microprocessor, which communicated with the sensors, activated corrective measures when the sensors showed charging was underway. This system provided engineering design data for future operational systems while demonstrating and validating active charge control principles and technology.[95]

Two significant programs at the Battlespace Environment Division address the complexities of ionospheric scintillation. Personnel there developed the Scintillation Network Decision

The Combined Release/Radiation Effects Satellite was configured for a rocket-launched, single-orbit mission in 1987. Its solar panels have been drawn in to fit them to the fairing for the rocket.

Aid (SCINDA), allowing the Air Force Space Command to predict satellite outages along the equator caused by naturally-occurring disruptions in the ionosphere. SCINDA collects data from a series of eight locations around the equator. Every 15 minutes the SCINDA system accesses the data from the worldwide network to create simple, three-color maps of areas on or within 20 degrees of the equator that might experience communications outages. These maps are automatically updated every 15 minutes. Because SCINDA also shows users areas that are not suffering from scintillation, operators may be able to maintain communications by easily adjusting an antenna.[96]

The second system, the Communication/ Navigation Outage Forecasting System (C/NOFS) satellite, is a joint effort by the Air Force Research Laboratory and the DoD Space Test Program. Described by the Research Laboratory as the forecasting system of the future, C/NOFS (which should be ready for launch and operations within a few years) features three innovative core elements. First, the satellite will carry a sensor suite into space to take in-situ and remote sensing ionospheric measurements. Second, ground sensors will be able to use C/NOFS data to augment theater coverage for scintillation specification. Third, models and other capabilities will be able to create tailored outage forecast maps for use by America's warfighters.[97]

Another solar weather event with importance for space- and ground-based systems is a coronal mass ejection (CME). Similar to a hurricane on Earth, a CME is the eruption of a huge bubble of plasma from the Sun's outer atmosphere, its corona. CMEs are the principal ways that the Sun ejects material and energy into the solar system; they typically speed through the universe at about 1,000,000 mph. A single CME carries more than 10 billion tons of hot, electrically charged gas into the solar system. As a CME moves away from the sun with a force of billions of tons of TNT, it creates a shock wave as it hits the slower solar wind. The force caused by a wave of CMEs in 1998 pushed the leading edge of Earth's magnetic field down to 15,300 miles; it normally stretches about 45,000 miles from Earth toward the Sun. In effect, satellites normally protected by the magnetosphere in geosynchronous orbit at 22,300 feet suddenly found themselves twisting in the solar wind. The satellites didn't move; their whole neighborhood in space moved away for a while.[98]

The origins of such forces and their results are of obvious concern and interest to the Air Force and others who operate equipment in space and on the Earth. The Battlespace Environment Division currently has an experiment in space, launched as a secondary payload on the Coriolis mission on 6 January 2003 from Vandenberg AFB, California, called the Solar Mass Ejection Imager (SMEI). Scientists and engineers from the Air Force Research Laboratory, the University of California at San Diego, and the University of Birmingham, United Kingdom, worked together to design and fabricate the SMEI experiment and system. The SMEI has the capability to use its all-sky camera to take images that should help space-weather forecasters take a giant step forward in improving their forecasts. By detecting CMEs directed at the Earth, SMEI will help protect space assets and maintain stable communications, both of which are of immense importance to the warfighter. The all-sky images taken from SMEI will also benefit astronomers and astrophysicists in understanding solar processes and detecting astronomical phenomena.[99] SMEI had already taken its first spectacular all-sky image of the sun within a month of liftoff.[100]

SCINDA, C/NOFS, and SMEI represent ongoing efforts by the Air Force to create, refine, and distribute space-weather information to the warfighter. As these systems are updated and modified, they will become the basis for more advanced systems to provide timely and accurate information to the nation's defenders. Advanced state-of-the-art systems may come from breakthroughs in computer technology, better instrumentation for measuring ionospheric scintillation, or some other event. The ongoing nature of this research underscores the constant contribution of scientific research and development to the growing importance of space to America.

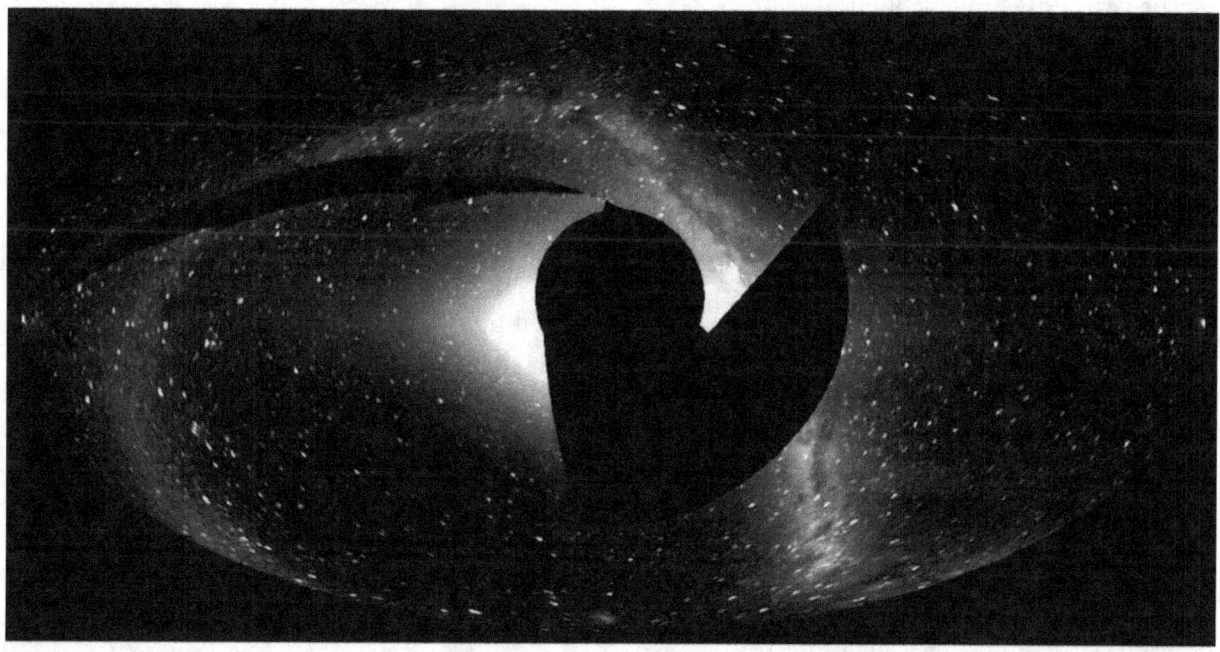

The Air Force Research Laboratory Space Vehicles Directorate's Solar Mass Ejection Imager launched on 6 January 2003. The SMEI team at the Directorate's Battlespace Environment Division shared its first all-sky image taken by the imager less than a month after launch (above). In March 2003, the team released an improved SMEI composite all-sky image (below).

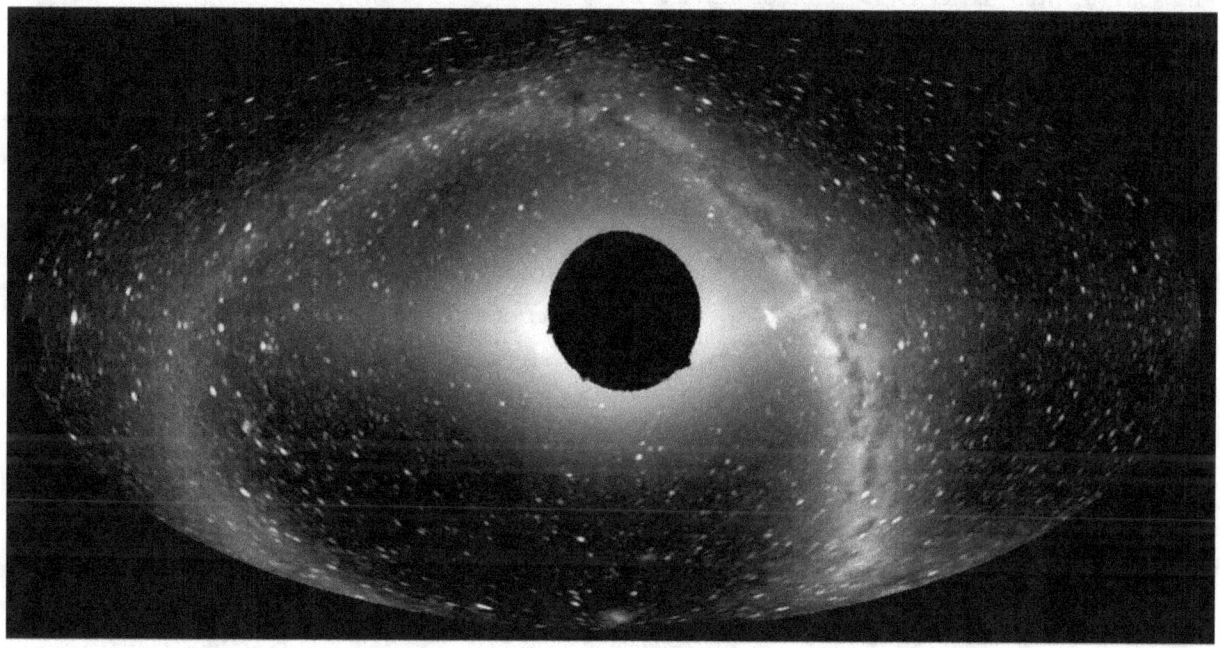

SUMMARY AND CONCLUSION

The year 1903 brought about several important events. Panama declared its independence from Colombia, allowing President Theodore Roosevelt to begin work on the Panama Canal the next year. The canal made navigation more convenient in terms of cost and time. It also allowed for quicker mobilization of the American fleet from the Pacific to the Atlantic in time of need, a lesson learned in 1898 with the voyage of the USS *Oregon* from the Pacific coast to Cuba in support of the American fleet during the Spanish-American War. The first narrative silent movie, *The Great Train Robbery*, premiered, changing the motion picture by its use of jump cuts and crosscuts, outdoor locations, and camera placement. In Milwaukee, William Harley and the Davidson Brothers Arthur, Walter, and William, culminated their work on "taking the work out of bicycling," and sold the first Harley-Davidson motorcycle. The motorcycle sported a one-cylinder gas combustion engine and was one of the three motorcycles the company manufactured that year and sold through its first dealership, located in Chicago. And Henry Ford and eleven other investors incorporated the Ford Motor Company and sold its first car, a Model A.[101]

Arguably the single event that most revolutionized the world that year was the 59-second, 852-foot first flight of a heavier-than-air machine at Kitty Hawk, North Carolina. Powered flight, and the aircraft and spacecraft we enjoy today, have come farther than any of the other technologies just mentioned. Motorcycles still carry only a few passengers, cars may get better mileage, and both may be more reliable than their 1903 counterparts. The Panama Canal still plays an important role in the global economy and America's defense, and the Canal also has determined ship design for some time, for passage through it is important for large and small vessels. Just as the national economy was once built upon the railroad and personal mobility increased with the appearance of the automobile, so has air transport of people and cargo made the world a closer-knit place.

Military operators, over the years, have come to rely on the tools created by Air Force scientists and engineers for understanding the space environment and space weather. Operational users have long enjoyed the benefits of state-of-the-art Air Force and Air Force–sponsored research and development. From 1971 through 1998, the Air Force operated the Defense Meteorological Satellite Program (DMSP), a series of satellites tasked "to generate terrestrial and space weather data for operational forces worldwide." DMSP-generated data is also furnished to the civilian community through the Department of Commerce via NOAA [National Oceanic and Atmospheric Administration]. Air Force researchers played a major role in the development of a new class of remote atmospheric sensing systems and transferred these to DMSP. Among other functions, these sensors solved problems with previous generations of atmospheric sounders and allowed DMSP to provide continuous atmospheric soundings at altitudes ranging from the earth's surface to 40 km [kilometers]. The Air Force Weather Agency uses the data collected in this manner to model the earth's winds and cloud movements and provide timely, improved forecasts to all DoD forces and operating locations as well as to civil weather forecasting agencies. In 1998, when the Air Force and the Department of Commerce agreed to transfer control of DMSP to a joint operational team, the Air Force contributed its scientific expertise directly to civilian users, a major contribution to the American people.[102]

This is a fine example of how research and development experts, operators, and the American people have benefited from the state-of-the-art research the Air Force conducted on the space environment and space weather. Starting with lessons learned during World War II and using captured German V–2 rockets, Air Force science and technology has grown apart from terrestrial meteorology. Given the longer history of global weather data collection and dissemination, space-weather research is relatively well advanced after fifty years, in comparison with the work of early weather scientists like Benjamin Franklin in the 1750s.

The Air Force has pursued space-related research from ground stations, aircraft, balloons, and outer space. Individually and collectively, these systems have quickly produced a series of unfolding and ever-changing definitions of what is state of the art in terms of our understanding of the space environment. As computing technology has matured since the 1940s, so, too, has the Air Force's ability to look into the vastness of space to discover what exactly faces the space systems of the future.

Since the discovery of the Van Allen radiation belts in 1958, experts with the Air Force have dedicated themselves to the pursuit of knowledge in the belief that space holds the answers for many phenomena on earth. They have discovered myriad events related to the sun that create CMEs, atmospheric scintillation,

and solar flares, and they have helped explain coronal holes, sunspots, and geomagnetic storms. All of these solar events can have dramatic impacts on man-made electronic systems on earth and in space. With the push of new technology pulling the development of new devices from black-and-white to color television to today's high-definition television; from local broadcasts to satellite dishes; from the first hand-held calculators in the 1970s to the first affordable home computers in the 1980s to today's cellular telephones, personal digital assistants, and Blackberry devices technology has been the hallmark of the post–World War II world. Military systems are likewise more dependent than ever upon consistent performance by electronics and computer hardware and software. Advanced work by the Air Force's best and brightest is making sure that in the future, as they have in the past, Americans can enjoy the prosperity that comes from a world at peace, a world that can remain in touch, no matter the time or weather.

The Air Force is well known for its advancements in air power and its important role in America's wars in the last sixty years. But the Air Force has also created an impressive history in space-related accomplishments in conjunction with civilian counterparts at NASA, the Department of Commerce, and the National Oceanic and Atmospheric Administration. The same is also true of Air Force research and development efforts aimed at understanding and forecasting the space environment and the myriad events that constitute space weather. As the world becomes more attached to its electronic devices, as power grids continue to grow and carry heavier burdens throughout the world, as satellites carry more and more of our economy, communications, and entertainment, and as space plays an increasingly more important role in our national security, an ability to predict the impacts of space weather on these systems is more important than ever. Space may be the next frontier, but without an understanding of the challenges in space and the impact of space weather on space- and ground-based systems, the promise of tomorrow cannot be attained.

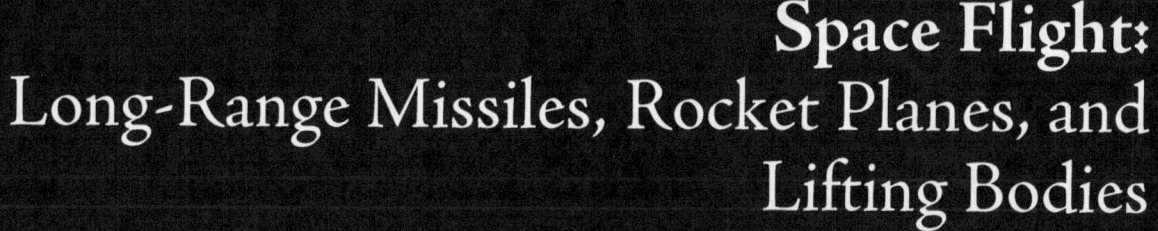

Space Flight:
Long-Range Missiles, Rocket Planes, and Lifting Bodies

Rick W. Sturdevant

John H. Darrah

Lieutenant General Forrest McCartney, USAF, Ret.

Abstract

Almost from their inception, Air Force long-range ballistic missiles—Atlas, Titan, and Thor,—which were designed to project thermonuclear warheads toward targets thousands of miles distant, provided the nation with a significant space launch capability. These Intercontinental Ballistic Missiles (ICBM) and Intermediate Range Ballistic Missiles (ICBMs) were transformed in the late 1950s into dependable space launchers and eventually into today's Evolved Expendable Launch Vehicles. The Air Force also contributed to the development of several powerful upper stages that significantly improved national space-launch capability.

Moving into the realm of reusable launch vehicles, the Air Force led, or participated in, development and testing of several manned aerospace vehicles. Cooperation between the National Aeronautics and Space Administration and the U.S. Air Force has been key to the advancement of reusable launch vehicle technology: from the X–15 and Dyna-Soar to lifting bodies and the Space Shuttle. Pursuit of a National Aerospace Plane in the early 1990s highlighted the utility of a military space plane as part of a new generation of reusable launch vehicles to replace the shuttle.

INTRODUCTION

Rocketry—the science related to non-air-breathing pyrotechnic devices, that is, missiles, propelled by hot gases—led to a technology applied initially to advanced warfare. At least seven centuries of development preceded efforts in the mid-1900s to build missiles able to traverse thousands of miles and deliver thermonuclear warheads against enemy targets. Almost simultaneously, it became apparent that long-range military rockets also might serve astronautics, that is, manned or unmanned navigation, exploration, and utilization of outer space. As an integral part of astronautics, rockets function as boosters for spacecraft.

Nearly sixty years ago, even before it became a separate service, the United States Air Force (USAF) undertook the design and acquisition of long-range missiles as weapon systems. By the 1960s, those missiles also had become the boosters that put American astronauts, satellites, and interplanetary probes into space. The vast majority of all American space launches from that time to the present, regardless of whether the payload belonged to National Aeronautics and Sapce Administration (NASA), a private company, a foreign nation, the USAF or another military service, or the once super-secret National Reconnaissance Office, have been achieved via USAF expendable boosters or their variants.

In addition to expendable boosters derived from long-range ballistic missiles   the intermediate-range Thor and intercontinental Atlas and Titan, the USAF has contributed significantly over the last half century to the development of powerful upper stages to propel spacecraft beyond low Earth orbit. The USAF also has worked extensively over many years on the design of hybrid (partially reusable) or fully reusable launch and powered-flight systems, even though these have been primarily under NASA's purview for purposes of manned space flight. Unquestionably, America's access to space has relied in large measure on launch capabilities provided in whole or part by the USAF.

LONG-RANGE MISSILES— EXPENDABLE BOOSTERS

American military interest in long-range missiles increased dramatically during the last years of World War II, when German V–2 rockets began raining down on London and other Allied targets. As early as 1944, the Jet Propulsion Laboratory in Pasadena, California, reported to the Army Air Forces on the feasibility of developing an American equivalent to the V–2 to deliver warheads over great distances. From that point through the early 1950s, however, military priorities and drastically diminished budgets prevented any sustained ballistic missile development program. Even a RAND report in early 1946 on the feasibility of an Earth-circling satellite, which suggested a secondary use for such missiles, failed to alter the situation. Not until the creation of smaller, thermonuclear warheads, combined with increasingly clear evidence of a long-range rocket program in the Soviet Union, did the USAF gain strong governmental backing for accelerated development of an intercontinental ballistic missile (ICBM). In 1954, the service established the Western Development Division under the leadership of Brigadier General Bernard A. Schriever to oversee ICBM acquisition. In due course, primarily to ensure that nothing competed with the ICBM program and secondarily to leverage similarities in technology, the USAF space program was transferred to the Western Development Division.

THOR-DELTA

Suspicious that the Russians were at least three years ahead of the United States in ballistic missile development and that accelerated procurement of an operational American ICBM involved considerable risk, the USAF opted through a 27 December 1955 General Operational Requirement to acquire an intermediate-range missile capability within one year. Douglas Aircraft delivered the first Thor intermediate-range missile on 26 October 1956. Powered by a single Rocketdyne MB–3 Block II engine burning RJ–1 (kerosene) and liquid oxygen, the Thor, which measured 56 feet high and 8 feet in diameter, boasted 150,000 foot-pounds of thrust at sea level. Gimbaling of the main engine controlled pitch and yaw, while two vernier engines controlled roll and pitch adjustment for the main engine. As a space launcher, Thor could loft 1,570 pounds. into low Earth orbit. When three Thiokol TX–33–52 (Sergeant) solid rocket boosters and an improved MB3 Block III main engine were installed in 1963 to produce the Thrust-Augmented Thor, it achieved 135,396 foot-pounds of thrust at liftoff and could place approximately 2,200 pounds into a 100-nautical mile orbit. Subsequent modifications to the original booster in the mid-1960s lengthened the tank by eleven feet to produce the Long-Tank Thrust-Augmented Thor.

A tally of Thor space launches includes many

noteworthy successes. Among the earliest were Discoverer I, the first polar-orbiting satellite, in February 1959; Explorer I, the first to photograph Earth from space, in August 1959; Pioneer 5, which entered solar orbit, in March 1960; Transit 1B, the first navigational satellite, in April 1960; Discoverer XIV, the first mission to return reconnaissance photographs under the Corona program, in August 1960; Courier 1B, the first active-repeater communication satellite, in October 1960; and Oscar 1, the first amateur radio operators' satellite, in December 1961. Thor became the mainstay for launching

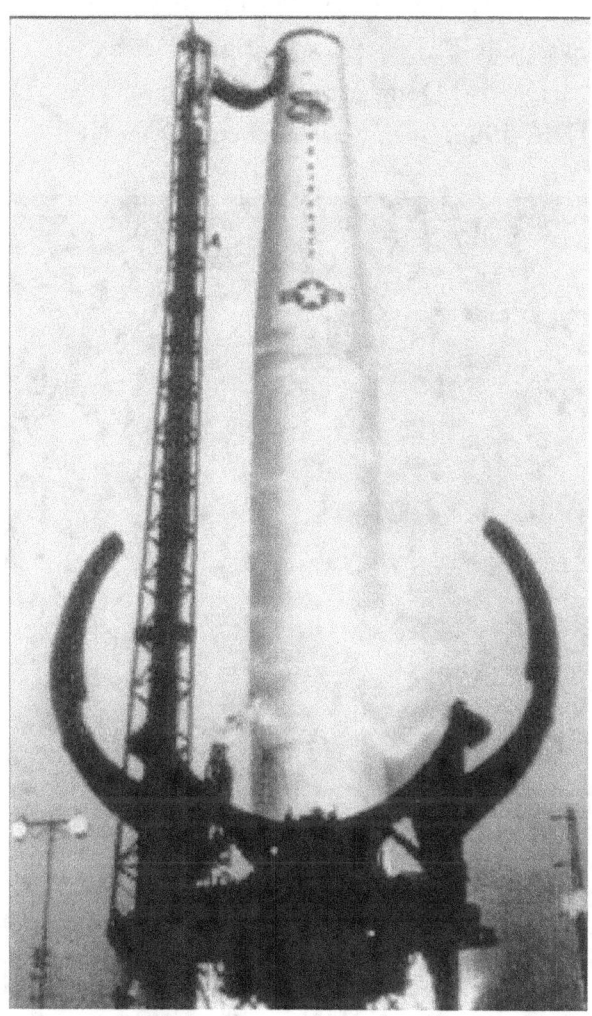

Figure 1. Thor IRBM.

153

Corona reconnaissance satellites as well as associated Defense Meteorological Satellite Program (DMSP) payloads, during the 1960s. When the last Thor-Delta was launched on 6 October 1981, the Thor booster had flown more than 500 times in various combinations.

Under a NASA contract in 1959, Douglas Aircraft combined Thor components with U.S. Navy Vanguard elements to deliver, in only eighteen months, the Delta launch vehicle, which evolved over time to meet a variety of requirements. After a series of launch disasters during 1985–1987, the USAF, needing a launcher primarily for Global Positioning System (GPS) satellites, contracted with McDonnell Douglas to develop the Delta II. By 1998, Boeing had derived a Delta III

Figure 2. Thor Space Launcher.

from the Delta II to fulfill requirements for a higher capacity commercial launcher, one that could double the weight a Delta II could place in a geo-transfer orbit. While the Delta III added elements to increase performance, it shared a production line and many common systems (including the RS–27A first-stage main engine) with the Delta II.

Like its Thor progenitor, Delta proved remarkably successful. Its first success came in August 1960 when Echo 1, the first passive communications relay satellite, entered orbit. Delta boosters also launched NASA Tiros weather satellites; the first Orbiting Solar Observatory in March 1962; AT&T's Telstar satellites; and the first geosynchronous satellite, Syncom 2, in July 1963. In addition to being the workhorse for GPS launches, Delta II has launched NATO communications satellites and nonmilitary communications and scientific payloads for the United States and several foreign nations. One of its more recent achievements involved the extremely successful Near Earth Asteroid Rendezvous spacecraft launched in February 1996.

ATLAS

Initiated by Convair as Project MX–774 for the Army Air Forces in 1945, the Atlas ICBM program received highest national priority in March 1955. In terms of production, testing, and construction, it soon became the largest and most complex program ever undertaken, surpassing even the Manhattan Project. Headed by Belgian-born Karel J. (Charlie) Bossart, Convair engineers used an innovative, weight-saving "balloon" construction in which the extremely thin, pressure-stabilized propellant

Figure 3. Atlas ICBM.

Figure 4. Atlas-Mercury Launcher.

tanks were integral to the airframe itself. Although the first launch of a developmental Atlas A occurred in June 1957, the first operational Atlas D ICBM launch did not come until April 1959. The stage-and-a-half Atlas D, dubbed by one author the "granddaddy" of all versions over the next 40 years, evolved along two branches: one contained the E- and F-model ICBMs; the other focused on space launchers, beginning with the LV–3 series. Because tailoring each Atlas to a specific space mission lengthened lead times and increased cost, the USAF awarded the Convair Division of General Dynamics a contract in 1962 to develop a standardized launch vehicle. Twenty-

five years later, after a national decision to remove commercial payloads from the Space Shuttle, the company opted on its own to build Atlas I. In May 1988, the USAF, needing a launcher primarily for Defense Satellite Communications System (DSCS) payloads, contracted with Convair to develop the Atlas II. Convair lengthened the Atlas I by nine feet to increase the amount of propellant the rocket could carry and employed an improved Rocketdyne MA–5A engine set that produced 414,400 foot-pounds of thrust. Further modifications, which replaced the multiple Rocketdyne engines with a single Russian-built NPO Energomash RD180 and drastically reduced the total part count to simplify construction, resulted in the Atlas IIAR. Later renamed the Atlas III, this launcher retained the pressure-stabilized structural concept of the original Atlas ICBM but dropped the stage-and-a-half design in favor of a single stage.

Atlas boosters achieved many prominent successes during the early years of space launch. First used in December 1958 for Project Score, the world's first communications satellite, Atlas boosters also lofted USAF Midas and Samos payloads beginning in 1960 and Vela nuclear detection satellites beginning in 1963. As the 1960s progressed, Atlas vehicles sent Ranger, Surveyor, and Lunar Orbiter spacecraft to the Moon in preparation for human expeditions. Mariner missions to Mercury, Venus, and Mars as well as Pioneer probes to Jupiter, Saturn, and Venus also employed the Atlas. NASA employed a man-rated version of the Atlas for every Mercury orbital flight beginning with John Glenn's mission on 20 February 1962. The last launch of an Atlas E placed a DMSP satellite in Sun-synchronous orbit on 24 March 1995. Atlas Is launched the Navy's Ultrahigh Frequency Follow-On (UFO) satellites and the National Oceanic and Atmospheric Administration's Geostationary Orbiting Environmental Satellites during the early 1990s. During the later 1990s, Atlas II took over the UFO launches, along with DSCS and various other kinds of communications payloads, both commercial and civil.

TITAN

Considering the urgent requirement for an operational ICBM and recognizing the cutting-edge technological and manufacturing risks associated with Atlas development, the USAF opted in October 1955 to contract with the Martin Company for a two-stage Titan ICBM. This parallel development program stimulated industrial competition to produce a weapon system in the shortest time. To reduce the risk of failure even further, both the Titan and Atlas programs employed separate associate contractors for each major subsystem (e.g., airframe and assembly, propulsion, guidance, computers, and nose cones), thereby improving the odds that critical components might be interchangeable between the two missiles. Over the longer term, this approach ensured expansion of research and development (R&D) firms and, in turn, increased production capabilities relevant to missiles and space.

Figure 5. Titan II ICBM.

Figure 6. Titan II SLV.

Figure 7. Titan IVB.

The Titan I ICBM, which first flew in February 1959, became the progenitor of a whole family of space launchers that culminated with the Titan IV heavy lifter, which first flew in June 1989. Rather than rely on incremental changes to Titan I, the USAF chose to alter significantly the configuration of the Titan II ICBM. The latter had a self-contained, all-inertial guidance system that allowed a "salvo" launch of the entire force, and storable, hypergolic propellants instead of cryogenic fuels enabled a more rapid operational response. Each Titan II could be stored in and launched from a hardened, underground silo, and a system involving two main thrusters allowed for "steering" the missile in a novel fashion. A much greater "throw weight" permitted delivery of heavier thermonuclear payloads anywhere on the globe.

Because problems and inefficiencies still existed in tailoring individual Titan IIs for specific space-launch missions, the USAF in 1961 initiated development of Titan III. A Titan II ICBM, structurally modified to handle increased payloads, served as the Titan III common core. Using that standard core in combination with solid-propellant, strap-on rockets, or in a stretched version, the USAF produced a variety of space launchers between 1965 and 1982: Titan IIIC, D, and E; 23B, C, D, and E; 24B; 33B; and 34B and D. From the Titan 34D, Martin Marietta derived a commercial Titan III that first flew in 1989. Meanwhile, deactivation of the Titan II ICBMs, which the USAF completed in 1987, made them available for refurbishment as space launch vehicles with modification of their guidance and attitude-control systems and addition of instrumentation and range safety destruct subsystems. The first Titan II space launch flight occurred in September 1988 from Vandenberg AFB, California. Finally, Titan IV assured access to space for heavy payloads in the wake of the 1986 Challenger disaster. Two seven-segment solid rocket motors attached alongside a stretched, strengthened version of the Titan III core vehicle enabled the USAF to place payloads of 31,000 pounds in polar orbit from the West Coast or 39,000 pounds into low Earth orbit from the East Coast. That capability meant the USAF, using a Centaur upper stage in conjunction with the Titan IV, could send as much as 10,000 pounds into geosynchronous orbit. With the successful emergence of Evolved Expendable Launch Vehicles (EELVs), however, Lockheed Martin staked its future on Atlas V and brought the Titan genealogy to an end. The last Titan IV is scheduled to launch a Defense Support Program satellite early in 2004.

Like Atlas, Titan compiled an illustrious record of successful space missions. NASA employed man-rated Titan II boosters for every Gemini orbital mission during the 1960s. During the same period, Titan IIIs launched Lincoln Experimental Satellites, as well as Initial Defense Communications Satellite Program payloads, Vela satellites, and reconnaissance platforms. In the wake of the Challenger disaster on 28 January 1986, Titan IV heavy lifters handled Milstar, Defense Support Program, and other defense-related payloads, while refurbished Titan II ICBMs furnished reliable space boosters for DMSP.

EVOLVED EXPENDABLE LAUNCH VEHICLE

When exploration of alternative space-launch systems during the 1980s and early 1990s ended without a clear national consensus, Congress tasked the Department of Defense (DoD) with identifying the most appropriate way to remedy launch deficiencies and reduce the rising cost of launch services. A team headed by Lieutenant General Thomas Moorman, vice commander of Air Force Space Command, responded in April 1994 with the Space Launch Modernization Study, which considered four options: sustain existing launch systems; evolve current expendable launch systems; develop a new expendable launch system; and develop a new reusable launcher system. Based on the Moorman study's recommendation, the National Space Transportation Policy, which President Clinton signed on 5 August 1994, directed the Secretary of Defense to implement the second option. There simply was not sufficient federal funding to develop a completely new expendable or reusable launch system, and high costs combined with inadequate flexibility weighed against simply sustaining the existing fleet. On 25 October 1994, Deputy Secretary of Defense John Deutsch signed the implementation plan for evolution of the current expendable launch vehicle fleet. The USAF had management responsibility for the EELV program, but, unlike in earlier booster development programs, this time significant cost-sharing between the corporations and the government was involved. In October 1998, the USAF awarded contracts to both Lockheed Martin and Boeing for EELV development and initial launch services. Innovative management by the USAF resulted in trimming the traditional seven-year acquisition cycle to four years.

The EELV fleet consists of two competing booster families - the Lockheed Martin Atlas V and the Boeing Delta IV - each blending new and mature technology, each based on a common booster core, and each deliverable in a variety of configurations (i.e., number and type of solid-propellant, strap-on boosters, as well as number of upper-stage engines and faring sizes) to meet military, civil, and commercial customer requirements through the year 2020. Atlas V is available in approximately fifteen mix-and-match variants, all of which rely on a common core that incorporates a reinforced first-stage structure powered by a single, Russian-built RD180 engine. Depending on the variant, Atlas V can place more than 18,000 pounds into a geo-transfer orbit or 13,000 pounds directly into a geostationary orbit. Delta IV employs a common-core first stage powered by a single Rocketdyne RS–68 engine and appears in at least five variants employing Delta II or III second stages along with strap-on boosters— either the solid-propellant, Graphite Epoxy Motor, or the liquid-propellant, common-core one. Depending on which variant is selected, the Delta IV can place between 9,000 and 29,000 pounds into a geo-transfer orbit. The inaugural flight of an Atlas V, carrying a Eutelsat Hotbird 6 payload, occurred on 21 August 2002, followed by the inaugural flight of a Delta IV, carrying a Eutelsat W5 payload, on 20 November 2002. Another milestone came on 10 March 2003 when only the third EELV ever launched, another Delta IV, successfully placed USAF DSCS payload in its target orbit.

MINUTEMAN

Unlike its predecessors, the Atlas and Titan, the solid-propellant Minuteman ICBM, which became operational in 1962, never evolved to a standard space-launch vehicle. Minuteman technology, nevertheless, enabled production of large, solid rocket boosters for Titan III and IV space-launch vehicles as well as even larger Solid Rocket Boosters for the Space Shuttle. Furthermore, Minuteman stages have been used in conjunction with commercial upper stages for satellite launches. Most of the credit for sustaining an interest in large, solid-propellant rockets during

Figure 8. Atlas V.

The Limitless Sky

the mid-1950s belongs to USAF Colonel Edward N. Hall, who touted their simplicity compared to liquid-propellant systems. Even before the USAF gained DoD approval for the Minuteman program in February 1958, Hall's efforts contributed substantially to the Navy's Polaris Submarine Launched Ballistic Missile. Minuteman development prompted the use of aluminum as a fuel additive, the formulation of more powerful propellants such as polybutadiene–acrylic–acid–acrylonitrile for the Minuteman I first stage, and carboxy terminated polybutadiene for the Minuteman II second stage. Although carboxy terminated polybutadiene represented a significant advance in binder technology, it was very costly and was

eclipsed in the late 1960s by an even better, less expensive polymer hydroxyl-terminated polybutadiene. Material and manufacturing innovations for nozzle throats, exit cones, and cases paved the way for later applications in large, solid-propellant space boosters.

While the Minuteman did not become a standard space booster in its own right, it did provide two stages for the hybrid, four-stage Minotaur, which Orbital Sciences Corporation produced under a September 1997 USAF contract. Officially known as the Orbital Suborbital Program Space Launch Vehicle, the Minotaur combined Minuteman II first and second stages with the second and third stages of Orbital's Pegasus XL small launch

Figure 9. Delta IV.

160

launch in July 2000 successfully placed the USAF MightySat 2.1 experimental payload in Sun-synchronous orbit. Minotaur missions were restricted to government and university payloads on the basis of national need.

POWERFUL UPPER STAGES

Although more attention generally has been given to large single or multistage boosters, some of the most advanced rocket research has focused on upper stages. Performing their tasks at relatively high altitudes, beyond the range of visible exhaust plumes, these relatively unsung heroes of the space age play a critical role in placing larger payloads into high Earth orbit or sending interplanetary spacecraft on their merry way across the heavenly expanse. Major upper stages generally depend little on booster or spacecraft subsystems for functions like navigation, telemetry, or control. Consequently, they represent a distinct category of propulsion systems, one to which USAF-sponsored R&D has contributed significantly over many years. Some of these upper stages (e.g., Agena, Centaur, and Inertial Upper Stage) contributed to the success of numerous civil and commercial launches as well as military, while others (e.g., Burner II and Transtage) generally saw defense-related service.

AGENA

Lockheed began working on the Agena, originally called Hustler because it used a Bell Aerospace engine by that name, in 1956. The USAF had contracted for such a vehicle to propel a WS–117 L satellite payload into orbit and, remaining attached, to supply the payload with power and directional pointing.

Figure 10. Minotaur.

vehicle. By using parts from 350 retired ICBMs, the USAF hoped to slash millions of dollars from the cost of launching military research satellites into low Earth orbit. The Minotaur could launch 1,400 pounds into low Earth orbit for an estimated cost per pound of $9,000 compared to $15,000 per pound using a Pegasus XL. In January 2000, the inaugural Minotaur launch successfully placed the Joint Air Force Academy–Weber State University Satellite (JAWSAT) in orbit. Also serving as a multi-payload adapter, JAWSAT carried four microsatellite payloads. One of the latter, OPAL, built by Stanford University students, subsequently released the smallest satellites ever placed in orbit. Built by The Aerospace Corporation, the Picosat 1A and 1B tethered satellites each measured approximately 4 inches by 3 inches by 1 inch. Another Minotaur

All Agena upper stages relied on a single engine, and all except the very first one, which burned JP–4, used storable, hypergolic propellants: inhibited red fuming nitric acid/unsymmetrical dimethyl hydrazine. The first model, Agena A, became available in February 1959. Atop a Thor booster launched from Vandenberg AFB, the Agena A could place roughly 1,700 pounds into low Earth, polar orbit. A structural redesign in October 1960 nearly doubled the propellant capacity, which allowed the Agena B version atop a Thor to place approximately 2,500 pounds into a similar low Earth, polar orbit. Upgrades to improve producibility, versatility, guidance accuracy, and on-orbit longevity led to the Agena D in June 1962. The latter in combination with a Titan IIIB could place approximately 7,000 pounds in polar orbit from Vandenberg. Both the last Agena A and the last Agena B used Atlas boosters in January 1961 and June 1966, respectively. Production of the Agena D ceased in 1982, and the last flight occurred in February 1987 on a Titan IIIB. During its last two decades of operation, the Agena D registered a success record of 100 percent during 131 missions. In every instance, the booster operated correctly, giving the upper stage an opportunity to perform.

CENTAUR

Although NASA generally receives credit for developing the Centaur, this liquid-propellant, high-energy upper stage originated in 1956 with Krafft Ehricke, an employee of Convair/Astronautics Division of General Dynamics. That company wrote a detailed proposal in 1957, which the USAF exhaustively reviewed. In January 1958, Air Research and Development Command proposed a program to the Air Staff to develop such an exploratory, experimental space vehicle. In August 1958, the Advanced Research Projects Agency formally established the Centaur program and assigned management responsibility to the USAF. The Centaur, powered by newly designed Pratt and Whitney RL–10 engines burning a liquid oxygen–liquid hydrogen mixture and flown atop Atlas or Titan boosters, would place heavy payloads into geosynchronous orbit. Relying on a balloon tank design similar to what Bossart used for the Atlas ICBM, Ehricke's Centaur nevertheless featured an innovative, technically demanding double-walled integral bulkhead to provide insulation between fuel and oxidizer. Although Centaur was the first space vehicle to burn liquid hydrogen, one could trace USAF pursuit of

Figure 11. Agena Upper Stage.

Figure 12. Centaur Upper Stage.

liquid hydrogen as a rocket fuel to a July 1945 contract with Ohio State University's Cryogenic Laboratory. Despite this longstanding interest, development of the necessary technology proved extremely challenging. The USAF transferred the Centaur program to NASA on 30 June 1959, but NASA continued to use the USAF contracts with General Dynamics until January 1962. Not until November 1963 did the first successful developmental flight occur, and it would be another three years beyond that before Centaur became operational.

BURNER II

In September 1965, the USAF announced development under a Boeing Aerospace contract of a new, low-cost upper stage Burner II that would be adaptable to almost any standard USAF booster and would be the smallest maneuverable upper stage in its inventory. Relying on a Thiokol STAR 37B solid-propellant motor for its main propulsion, the Burner II first flew on a Thor in September 1966. The Burner II permitted direct injection, Hohmann transfer perigee burns, and bi-elliptic transfer burns. When used with a Thor for direct ascent from Vandenberg, the Burner II could put a payload weighing approximately 125 pounds into a 1,000-mile circular orbit. Desiring to add an apogee circularization capability, the USAF requested in June 1969 that Boeing modify the Burner II. By adding a second stage powered by a Thiokol STAR 26B solid-propellant motor, the Burner IIA nearly doubled the on-orbit capability of Burner II, thereby improving the ability to circularize payloads in operational orbits and allowing changes in the orbital plane. Placement of flight-

proven Burner II equipment and subsystems (e.g., guidance and flight control, reaction control, electrical and telemetry subsystems) in the apogee kick stage allowed jettisoning of the larger, first-stage motor after burnout, which enhanced overall performance of the upper stage. Atop an Atlas 3A, the Burner IIA could put roughly 500 pounds into geosynchronous orbit. The first Burner IIA flew in October 1971. Used primarily to launch DMSP satellites, the last successful Burner II flight occurred in June 1979, after which the USAF launched those satellites on higher-performance boosters that did not require an upper stage.

INERTIAL UPPER STAGE

When Vice President Spiro Agnew's Space Task Group learned in 1969 that NASA was projecting another 15 years before its space tug would be operational, the group favored rapid development of an Interim Upper Stage (IUS) to bridge the gap. Despite the perceived urgency for producing an operational upper stage, official records reflect little progress until the USAF assumed, in a May 1973 memorandum on the DoD use of the Space Shuttle, military development of the IUS. In October 1973, NASA agreed that the USAF should oversee IUS development, and in September 1976 Boeing Aerospace received the validation-phase contract for a new two-stage, solid-propellant IUS, one that would be compatible with Space Shuttle designs and pose less danger to the flight crew than a cryogenic system would. Introduction of a USAF requirement that the IUS also be compatible with the Titan 34D booster, the largest expendable booster then available, compelled Boeing to abandon

163

its original intention of simply designing a "growth" version of the existing Burner II. Cancellation of the NASA space tug program at the end of 1977 further refocused the IUS program and resulted in its name being changed to Inertial Upper Stage, a reference to its guidance-control technique. As full-scale development proceeded after April 1978, numerous technical difficulties in the propulsion subsystem, software, and avionics, as well as programmatic changes, shifting specifications, and a weight-growth problem, threatened to scuttle the entire program.

The resulting operational IUS, which first flew in October 1982, proved to be unique in several ways. It had a first-stage motor that could maintain continuous thrust longer than any other solid-propellant upper stage (up to 2.5 minutes); it was more functionally redundant than any other upper stage; and it was the only one that ground controllers could command during flight. Although its structure could support an 8,000-pound payload, Boeing studies suggested that modifications would allow it to handle as much as 16,000 pounds The IUS provided the first upper-stage guidance for the Titan 34D, and it also became the first upper stage used on both the Titan IVA and B.

TRANSTAGE

Martin Marietta, working under a USAF contract in the early 1960s, developed the Transtage for the Titan III. This particular upper stage underwent its first developmental flight test on a Titan IIIA in September 1964, graduating to a Titan IIIC on its fifth R&D flight in June 1965. In November 1970, the Titan 23C became the booster of choice for Transtage, and it remained so through another twenty-one missions over twelve years. From January 1984 to its final flight in September 1989, the Transtage rode atop Titan 34D boosters. Relying on twin Aerojet AJ10–138 or AJ101–38A engines capable of multiple starts in space, the Transtage atop a Titan 34D could deliver more than two tons to geosynchronous orbit. Its control module, which used an inertial guidance system to handle the entire Titan vehicle during flight, could be modified to separate from the propulsion module and remain with the spacecraft to meet attitude-control and

Figure 13. Inertial Upper Stage.

Figure 14. Transtage.

maneuvering requirements during the lifetime of the payload. The ability of a single Transtage to place several satellites in different orbits also proved valuable and made this upper stage a direct predecessor of the Multiple Independently Targetable Reentry Vehicle "bus" for propelling multiple small, thermonuclear weapons launched on a single ICBM to various targets.

REUSABLE LAUNCH SYSTEMS: ROCKET PLANES AND LIFTING BODIES

The USAF quest for reusable launch vehicles, manned or unmanned, began at the same time as its pursuit of long-range, surface-to-surface strategic missiles. As early as April 1946, the Army Air Forces guided missile program included Project MX–770, which involved a one-year contract with North American Aviation (NAA) for the study and design of a supersonic, winged rocket, the Navaho, having a range of 175 to 500 miles. Engine testing began with a surplus, liquid-propellant unit from Aerojet General that developed only 1,000 pounds of thrust. By June 1946, however, NAA proposed a two-phase engine development plan that involved refurbishing and testing a complete German V–2 propulsion system and subsequently redesigning it to meet American engineering standards and production methods. The company, drawing on the V–2 design but incorporating a number of improvements, added a third phase in early 1947 to produce a new engine. Even as the USAF was achieving independence in September 1947, NAA was

Figure 15. Navaho.

beginning preliminary design work on the Phase III engine. A major simplification, which translated into significant weight reduction without loss of thrust, was the replacement of multiple liquid-oxygen lines, each line feeding an individual injector, with a single injector plate that resembled a showerhead. This feature, applicable to larger engines with greater thrust than the 56,000 pounds produced by the V–2, became especially important when the USAF instructed NAA, in February 1948, to stretch the Navaho's range beyond 1,000 miles.

When the USAF doubled the range requirement, the original boost–glide approach to trajectory proved inadequate. The USAF proposed adding ramjet propulsion to the initial rocket boost to lengthen the time of supersonic cruise. The twin engines mounted on vertical tailfins, along with the necessary fuel supply, increased the total weight of the Navaho, which compelled modifying the Phase III rocket engine's thrust from 56,000 pounds to 75,000 pounds. Testing began in November 1949 and continued into March

1951, when engineers solved the vexing problem of combustion instabilities in the engine's thrust chamber. By then, progression of the Cold War and the Korean conflict had driven the USAF to extend the high-speed, pilotless aircraft's ultimate range requirement to 5,500 nautical miles. This, in turn, led to a fundamental design change that placed the rocket engines and ramjets in separate vehicles, making the Navaho a two-stage cruise missile. The ramjet-powered second stage, with its nuclear payload, would ride piggyback on the rocket-powered first stage to an altitude of 58,000 feet and a velocity of Mach 3, when the first stage would separate and return to base for reuse.

Although canceled by the USAF in 1957, the Navaho program contributed substantially to the future of American spaceflight. The missile's inertial guidance system found its way into nuclear submarines and Navy attack aircraft as well as into Hound Dog and Minuteman missiles. More importantly, the Navaho Phase III engine produced by the Rocketdyne division of NAA, along with those

Figure 16. X–15 Research Plane.

166

built in parallel by Aerojet General, became the basis for the propulsion systems in the Atlas, Titan, and Thor launchers, as well as in the Army's Jupiter. Further improvements increased the thrust of this basic engine design to 205,000 pounds. NASA clustered eight of these engines to produce the 1,600,000 pounds of thrust in the Saturn I and I-B boosters used so successfully in the Apollo and Skylab programs. In addition to hardware, the Navaho program enriched the aerospace research and production capabilities of NAA. The same NAA engineers who oversaw pioneering breakthroughs for the Navaho presided over such production triumphs as the Saturn V's main engines and the hydrogen-fueled engine that powered the Moon rocket's upper stages.

X–15

Not surprisingly, the USAF found the concept of an aerospace plane very attractive at an early date. If the Bell X–1 and X–2 programs were not direct precursors, they were at least harbingers of the X–15 program. The latter originated with the National Advisory Committee for Aeronautics (NACA), which decided in June 1952 that it should explore flight characteristics of atmospheric and exo-atmospheric designs capable of achieving velocities of Mach 4 to 10 and altitudes of 12 to 50 miles. In December 1954, the USAF and the Navy had joined NACA in forming a Research Airplane Committee, which oversaw the lengthy, complicated process of selecting a prime contractor, NAA, to build the experimental aerospace craft. Flights of the X–15 commenced in 1959 and ended in October 1968, after a total of 199 missions, 89 with

USAF pilots. During that decade of testing, USAF Captain Robert M. White accomplished the first astronaut wings flight by piloting the craft to an altitude of 314,750 feet in July 1962, and USAF Captain Joseph H. Engle became the only pilot to qualify for astronaut wings three times when he took the X–15 to 266,500 feet in October 1965. Another USAF pilot, Captain William J. Knight, set an unofficial world absolute speed record of Mach 6.70 in the X–15 in October 1967. Captain White was also at the controls in May 1960 when the U.S. received and recorded physiological data onboard an aircraft for the first time.

The contributions and spinoffs from the X–15 program were enormous. In October 1968, NASA engineer John Becker compiled an abbreviated list of 22 accomplishments that included the development of the first large, restartable, man-rated, throttleable rocket engine; first application of hypersonic theory and wind-tunnel work to an actual flight vehicle; first use of reaction controls in space, with successful transition from aerodynamic controls to reaction controls and a return to aerodynamic control; first reusable superalloy structure capable of withstanding hypersonic reentry temperatures and thermal gradients; first application of energy-management techniques; development of practical boostglide pilot displays; development of the first practical, single-piece, full-pressure suit for pilot protection in space; demonstration of a pilot's ability to function in a weightless environment and control a rocket-boosted aerospace craft during exo-atmospheric flight; and the first demonstration of piloted, lifting atmospheric reentry. Becker also mentioned that engineers studied hypersonic acoustic measurements from the

X–15 flights to define insulation and structural design requirements for the Mercury spacecraft.

A later listing of X–15 program accomplishments by Captain Ronald Boston of the USAF Academy History Department expanded Becker's assessment of the flight research program by discussing follow-on experiments that began around 1963 in the physical sciences, space navigation, reconnaissance, and advanced aerodynamics. Carrying instruments above the attenuating effects of Earth's atmosphere, the X–15 contributed to the physical sciences by achievements in photometric analysis of the ultraviolet brightness of several stars to determine their material composition; measurement of the atmospheric density to profile seasonal variations; and the first direct measurement of the Sun's irradiance from above the atmosphere, resulting in a revaluation of the solar constant of radiation that, in consequence, proved useful for designing thermal protection for spacecraft. The space navigation experiments aboard the X–15 furnished information about Earth's infrared horizon-radiance profile, which was used in attitude-referencing systems for orbiting spacecraft, and they collected data on the radiation characteristics of the daytime sky, which was applied in an automatic, electro-optical star trackingî system used on high-altitude reconnaissance aircraft and in satellite-positioning systems. Ultraviolet and infrared sensors aboard the X–15 tested the feasibility and relative efficacy of using those parts of the spectrum to detect and characterize the exhaust plumes of long-range missiles, which aided the development of satellite systems for missile-warning. Finally, the X–15 program spawned an enduring Mach 8 hypersonic ramjet engine project, even though X–15 flights ended before a prototype hypersonic ramjet engine could be delivered for actual flight tests.

X–20 DYNA-SOAR

Advocacy for USAF development of a fully reusable space vehicle based on the boostglide principle began with Walter Dornberger, a former general who headed Germany's military rocket program during World War II. During that period, he had become familiar with the work of Eugen Sänger and Irene Bredt, who collaborated on designing an antipodal bomber that would be boosted to orbital velocity, skip on the atmosphere, deliver its payload to the opposite side of the globe, and glide back to a friendly base for reuse. After the war, he had worked on missiles for the USAF at Wright-Patterson AFB, Ohio, before becoming a consultant to Bell Aircraft, where he and Kraft Ehricke, another German rocket scientist who had been brought to the United States through Project Paperclip, rejuvenated the Sänger-Bredt design study. In April 1952, Bell and Wright Air Development Center undertook joint

Figure 17. Preparing scale model of Dyna-Soar for wind tunnel test at Arnold Engineering Development Center.

development of a manned bomber and space reconnaissance vehicle called BoMi, which would be launched by a two-stage rocket to altitudes of 100,000 feet or higher and operate at velocities exceeding Mach 4. Through a series of USAF contracts with Bell, the BoMi project continued for another four years. During March 1956, however, the USAF channeled the development effort specifically toward a piloted, high-altitude reconnaissance system code-named Brass Bell. Before year's end, the contractor had designed a two-stage system powered by Atlas-type rocket engines and capable of reaching an altitude of 170,000 feet at a velocity of more than 13,200 mph with a range of 5,500 nautical miles.

Meanwhile, the USAF continued its pursuit of a manned, hypersonic, boostglide bomber by issuing study contracts to six aerospace companies Boeing, Republic, McDonnell, Convair, Douglas, and NAA for what became known in June 1956 as RoBo, a vehicle capable of reaching an orbital altitude of 300,000 feet and a velocity of 15,000 mph. In November, the USAF established a manned, glide-rocket research program known as HYWARDS (Hypersonic Weapons Research and Development System) to collect data on problems ranging from aerodynamics and structure to components and human factors. Faced with funding difficulties and the challenge posed by the Soviet Union's launch of Sputnik, the USAF opted in early October 1957 to consolidate HYWARDS, Brass Bell, and RoBo into a single Dyna-Soar (Dynamic Soaring) development program. Joined by the newly created NASA, the USAF contracted with Boeing in November 1959 to build Dyna-Soar based on a single-orbit operational principle.

By autumn 1961, however, a USAF decision to make Dyna-Soar a multiorbit vehicle forced Boeing to add a more sophisticated guidance system, improve the reliability of various subsystems, and add a retro-fire system for deorbiting the craft. Given the increasingly obvious experimental character of Dyna-Soar, it received the designation X–20 in June 1962. Plagued sporadically by less than wholehearted support from both within and outside USAF circles, as well as by perceptions of redundancy once NASA defined its Gemini program, the X–20 effort finally was canceled by Secretary of Defense Robert McNamara in December 1963.

Although Dyna-Soar never actually flew, its service as a testbed for numerous advanced technologies contributed much to the science of high-Mach flight. Advances in guidance-system technology proved important to the X–15 and later programs. Progress in aerodynamics, structures, and materials technology, much of it based on more than 14,000 hours of wind-tunnel tests, had significant worth for nearly two decades of subsequent lifting-body research.

Figure 18. X–20 Dyna-Soar.

When Rockwell International began design work leading to the Space Shuttle, the single most important U.S. database on fundamental reentry heating-dynamics technology came from the X–20 program. The Ren 41 high-temperature nickel alloy developed for X–20 heat shielding reappeared during the 1970s in Boeing's Reusable Aerodynamic Space Vehicle, a project in which the USAF invested several million dollars for development of a rocket-powered spacecraft that would have operated much like an aircraft in terms of takeoff and landing. Two classified, follow-on industrial studies of single-stage-to-orbit technologies, both sponsored by the USAF during the 1980s, suggested the feasibility of the reusable space vehicle for a variety of military applications, such as reconnaissance, rapid satellite replacement, and general space defense, but the USAF opted to drop the project and, instead, supported development of a National Aerospace Plane with air-breathing jet engines.

LIFTING BODIES—WINGLESS AND WINGED ROCKETS

During the late 1950s, NACA (and its successor NASA) as well as the USAF undertook research into lifting bodies—wingless, blunt-bodied craft that could maneuver at hypersonic velocities and reenter Earth's atmosphere as gliders. In 1959, the USAF initiated the minimal-cost ASSET (Aerothermodynamic/elastic Structural Systems Environmental Tests) project, which launched heavily instrumented, hypersonic glider models from Cape Canaveral on Thor and Thor-Delta boosters to investigate reentry from space at near-orbital speeds. ASSET added significantly

to the technology base for future manned reentry systems, especially regarding such issues as practical fabrication of refractory metals, coating processes, reaction-control system design, and the phenomenon of communications blackout during reentry. Furthermore, ASSET paved the way by early 1964 for an expanded Spacecraft Technology and Advanced Reentry Tests program with two parts: PRIME (Precision Recovery Including Maneuvering Entry), involving hypersonic boostglide tests using the SV–5D/X–23A unpiloted lifting body built by Martin Marietta; and Piloted Low-speed Tests, using the SV–5P/X–24A manned, rocket-powered lifting body, also built by Martin Marietta. Launched from Vandenberg AFB on Atlas boosters during 1966–1967, three X–23A PRIME flights entailed pioneering work in ablative materials and internal steam-cooling, accomplished the first cross-range maneuvering

Figure 19. X–24A (above) and X–24B (below) Lifting Bodies.

of a spacecraft, demonstrated accurate guidance to a selected recovery point, and supported the concept of a reusable spacecraft.

Jointly sponsored by the USAF and NASA, the X–24A/B project gained momentum with USAF approval for Martin Marietta to begin construction of the first and only X–24A, powered by a single Thiokol XLR11–RM13 rocket engine, in March 1966. Between April 1969 and June 1971, two USAF pilots and one NASA pilot completed 28 flights (9 were glides, 18 were powered, and 1 was a glide in middle of a powered flight) in the X–24A, which was dropped from a B52. A flight in October 1970 provided the first demonstration that an unpowered spacecraft could perform a shuttle-type approach and land on a conventional runway. In 1972, the USAF directed Martin Marietta in Denver, Colorado, to modify the X–24A shell to a more streamlined X–24B (basically giving it a new, pointed nose), which would allow testing of an aerodynamic configuration providing better maneuvering capability during reentry. Pilots flew the X24B thirty-six times between August 1973 and November 1975, twice landing successfully on a concrete, strip-type runway at Edwards AFB, California. This gave NASA engineers the confidence they needed to plan similar landings for the Space Shuttle, which was by then well into the design phase. Not surprisingly, that design owed much to the hypersonic flight research conducted by the USAF and NASA.

NATIONAL AEROSPACE PLANE

An aerospace plane—one powered by an air-breathing engine, capable of horizontal takeoff from a conventional runway, achieving orbital velocity, maneuvering in space, reentering the atmosphere, and landing conventionally — appealed to USAF officers. Indeed, the USAF had begun pursuing this concept in the late 1950s and, before funding ceased in the early 1960s, had shifted from a single-stage-to-orbit design toward a two-stage-to-orbit concept not unlike the later Space Shuttle. Before it died, however, this initial effort to develop an aerospace plane spawned potentially useful research on air collection enrichment systems, Mach 8 subsonic combustion ramjets, the Liquid Air Collection Engine System, scramjets, and advanced turbo-ramjets or turbo-accelerators. When the USAF Trans-Atmospheric Vehicle program began studying shuttle replacement in 1982, both air-breathing jet engines and rockets received serious consideration. Because several other organizations were pursuing similar work, it seemed efficacious by late 1985 to form a single National Aerospace Plane Program jointly sponsored and funded by the USAF, Navy, NASA, Advanced Research Projects Agency, and Strategic Defense Initiative Organization, with the USAF responsible for overall management. Despite the obvious military leadership and potential defense-related applications of the National Aerospace Plane, the Reagan administration touted its peaceful uses and envisioned the National Aerospace Plane as pioneering hypersonic commercial flights. Enthusiasts viewed it as the potential progenitor of all manned space transportation systems after the Space Shuttle.

Plans called for building the National Aerospace Plane flight-test vehicle, which received the designation X–30A, after validation of the high-risk technology associated

Figure 20. National Aerospace Plane.

with a Mach 25 aerospace craft capable of taking off and landing like a conventional aircraft. Development of airframes, materials, subsystems, and powerplants suitable for repetitive hypersonic flights posed a daunting challenge, but participating government agencies and private companies made significant technical progress before budget difficulties and the end of the Cold War led to cancellation of the National Aerospace Plane program in 1994. With respect to propulsion, the program registered several noteworthy accomplishments: thrust above Mach 8 was directly measured for the first time; large-scale ramjets were tested to Mach 8, large-scale scramjets were tested to Mach 16, and small-scale scramjets were tested to Mach 18; the inlet and combustor were tested to Mach 18 for the first time; the production, transfer, storage, and transportation of slush hydrogen were demonstrated; and the newly developed,

cryogenic 2D integrated fuel tank was tested. Exceptional progress also occurred regarding coated carbon-carbon composites, titanium aluminides, advanced metal matrix composites, copper niobium, and beryllium fiber material. Fabrication of advanced composite materials substantiated the capability to develop leading edges that could survive under near-flight conditions. Finally, the program yielded designs for integration of engine and airframe applicable to future spaceplane work, including a rounded-nose lifting-body concept. Some of the most valuable knowledge gained from the National Aerospace Plane program involved discovery of what would not work, in terms of both management structure and hardware.

With the National Aerospace Plane program scrapped, NASA sought to regain momentum through a cooperative agreement with Boeing to produce an X–37 reusable, orbital spaceplane. The unmanned, autonomously operated X–37 would provide a testbed for 30 to 40 airframe, propulsion, and operating technologies that might significantly reduce the cost of space transportation. Concurrently, the USAF financed a subscale version the X–40A Space Maneuver Vehicle to test the low-speed atmospheric flight dynamics of the X–37. Unveiled in September 1997, the graphite-epoxy and aluminum maneuver vehicle successfully completed a series of drop tests beginning in August 1998. Using an integrated GPS and inertial guidance system, the test vehicle autonomously acquired the runway in a simulated return-from-orbit and landed under its own power, similar to how a conventional aircraft would, thereby clearing the way for drop-testing the unpowered, full-scale X–37. A large portion of funding for the Space Maneuver Vehicle went into propulsion technology, with Aerojet receiving more than $10 million in May 2001 to develop a nontoxic, hydrogen peroxidebased system. By April 2002, the company had designed a revolutionary tri-fluid propellant injector for the Advanced Reusable Rocket Engine, and it was making progress in other risk areas, for example, in catalyst beds, thrust-chamber design, and turbine materials. As a reusable upper stage, the Space Maneuver Vehicle could be launched on several different boosters, but it was designed primarily as a key component of the Space Operations Vehicle system architecture, which relies on a reusable first stage and orbital maneuvering capability to accomplish a variety of military missions.

SUMMARY

Even as the nation contemplated the recent Columbia disaster and the implications for space flight, it continued to rely largely on expendable launch vehicles that originated from USAF-sponsored R&D programs in the late 1940s and 1950s. As requirements for more powerful launchers, more often than not defense-related, emerged during the 1960s and 1970s, the USAF contracted with industry to upgrade the existing Thor, Atlas, and Titan technology. When the Challenger accident compelled the United States to rethink its designation of the Space Shuttle as its sole means of space launch, USAF officials had already been considering how to revitalize

the production lines for expendable booster. To meet defense-related demands for medium and heavy lift, respectively, the USAF procured the Delta II and the Titan IV. With the end of the Cold War and the imposition of fiscal constraints on military spending, the USAF initiated acquisition of a launch vehicle to reduce significantly the cost of sending into orbit small, experimental payloads of national interest. The same fiscal constraints led to a national decision to continue the evolution of existing, highly reliable, expendable launchers rather than to embark on development of something entirely new at the beginning of the twenty-first century. Responsibility for managing that evolution fell squarely upon the USAF, which devised innovative partnering arrangements with industry to share the cost of acquiring EELVs.

As for reusable launch systems, USAF scientists and engineers considered them even before the establishment of the USAF as a separate military department. Working with the NACA and its successor, NASA, the USAF contributed extensively to the nation's understanding of the dynamics of hypersonic flight and to its technological base for producing the actual materials and subsystems needed for a space plane. Competing concepts of manned versus unmanned reusable vehicles generated reams of design studies over more than five decades and sparked countless hours of discussion about the relative advantages of one approach over the other. From the X–15, Dyna-Soar, and lifting bodies through Space Shuttle development and National Aerospace Plane, the USAF and NASA partnered to promote piloted, reusable systems for military, civil, and commercial purposes. Over time, however, the

USAF found itself reverting to the original notion of an unpiloted, reusable space plane, equipped with propulsion systems allowing it to operate efficiently through the atmosphere, into space, and, on its return, be capable of landing autonomously, like an aircraft, on a conventional, concrete runway. While NASA held primary managerial responsibility for reusable launch systems, the USAF and NASA worked diligently at the end of the twentieth century to improve their partnering arrangements and, thereby, avoid costly duplication of R&D efforts.

Military Satellite Communications:
From Concept to Reality

Harry L. Van Trees

Lieutenant General Harry D. Raduege, USAF

Rick W. Sturdevant

Ronald E. Thompson

Abstract

Military satellite communications have had a major impact on the success of recent military operations, such as Iraqi Freedom. Following a brief review of basic satellite characteristics—including orbital locations, frequency bands, bandwidth considerations, and terminal characteristics—the evolution of military satellite systems from a paper concept in 1945 to the sophisticated systems today.

Also discussed are technological developments that enabled the implementation of the Defense Satellite Communications System, Milstar, and Ultrahigh Frequency Follow-on systems, and various commercial adjuncts. Although some specific elements, such as multiple-beam antennas, phased arrays, signal design, and onboard switching are discussed, the technology focus is at the systems level. Specific operational successes that relied on military satellite communications are described. The technology developments required to support the evolving operational concepts are outlined.

INTRODUCTION

Military satellite communications have played a major role in the success of recent military operations. From Operations Desert Shield and Desert Storm to Iraqi Freedom, the precise coordination of assets would have been impossible without satellite communications (SATCOM). The ability to implement the sensor-to-shooter loops in near real time relied heavily on satellites.

This paper has three intertwined discussions: history, technology, and operations. We examine the history of SATCOM from a paper concept, appearing in science fiction articles in 1945, to the sophisticated, reliable systems of today. We focus on the key historical events that helped determine the evolution of SATCOM. In parallel, we discuss the key technology developments that have enabled the growth from the small (INTELSAT) I, Early Bird, launched in 1965, which carried 240 voice circuits, to the current large satellites, which carry massive amounts of voice and data traffic. The third goal of the paper is to demonstrate how SATCOM have affected military operations. Most of the new military operational capabilities in the information age rely heavily on SATCOM.

It is important to note that we use the term "military satellite communications," or MILSATCOM, to mean satellite communications for military purposes. The emphasis is on military satellite systems, but we will also discuss the usage of commercial satellite systems for military applications. The current Department of Defense (DoD) MILSATCOM system consists of five subsystems:

- Defense Satellite Communications System (DSCS)
- Milstar
- Ultrahigh frequency (UHF)
- Global Broadcast System (GBS)
- Commercial adjuncts

The evolution of each of these subsystems is described, including their configuration and their impact on recent military operations.

In an appendix to this paper, we review several satellite system fundamentals that will be helpful in understanding some of the information presented here. Readers not familiar with satellites may want to read this material before continuing with the paper

EARLY HISTORY OF SATCOM

The concept of using artificial, Earth-circling satellites for worldwide communications emerged near the end of World War II, a conflict that demonstrated the fundamental need for electronically transmitting military information over longer distances, in greater quantities, with more reliability and higher security than had ever been required before. In the February 1945 issue of the British technical journal *Wireless World*, science-fiction writer Arthur C. Clarke published a letter speculating on how three satellites positioned 120 degrees apart in geosynchronous orbit could relay television and microwave signals worldwide. Although theorists like Konstantin Tsiolkovsky (1911), Hermann Oberth (1923), and Hermann Potocnik (1929; pseudonym Hermann Noordung) wrote about space stations in geostationary orbits, and the latter two even speculated on the use of such stations as communication platforms, Clarke was the first to spell out essentially all the technical details for modern geosynchronous communications satellites. After circulating his ideas privately in a paper titled "The Space Station: Its Radio Applications," Clarke presented a more refined technical analysis of the orbital geometry and communications links in the October 1945 *Wireless World* article, "Extra-Terrestrial Relays."

Others soon echoed Clarke's basic notion. In a May 1946 report titled "Preliminary Design of an Experimental World-Circling Spaceship," Project RAND engineers at the Douglas Aircraft Company plant in Santa Monica, California, told the U.S. Army Air Forces that satellites could significantly improve the reliability of long-range communications and might spawn a multibillion-dollar commercial market. Subsequent RAND studies by James E. Lipp, in February 1947, and Richard S. Wehner, in July 1949, further developed the concept of geostationary communications satellites located above the equator. Eric Burgess further analyzed the possibility of using geosynchronous orbits for SATCOM and described in detail a potential satellite configuration in the September 1949 issue of *Aeronautics*. Writing under the pseudonym J.J. Coupling in *Amazing Science Fiction*, John R. Pierce of AT&T's Bell Telephone Laboratories suggested a communications satellite system in March 1952. He became one of the first people outside defense-related circles to evaluate, systematically, technical options and financial prospects for SATCOM. In a 1954 speech and 1955 article, Pierce assessed the utility of passive "reflector" and active "repeater" satellites at various orbital altitudes. If Arthur Clarke developed the theory of geostationary SATCOM, John Pierce and his team at Bell Labs pioneered and improved much of the hardware, for example, traveling-wave tubes and low-noise amplifiers that transformed theory into actual working systems.

The USSR's launch in 1957 of Sputnik I, which transmitted an electronic signal back to Earth simply for tracking purposes, sparked serious efforts by the United States to develop SATCOM for military, civil, and commercial use. SCORE (Signal Communication by Orbiting Relay Equipment), developed by the Advanced Research Projects Agency (ARPA) and launched by the Air Force in December 1958, became the world's first active communications satellite. During its twelve-day operational lifespan, which terminated when the battery failed, SCORE received messages from a ground station and stored them on a tape recorder for transmission back to Earth. The U.S. Army's Courier satellite, launched in October 1960, operated on much the same principles as SCORE, that is, store-and-dump with the use of onboard tape recorders, but Courier carried solar cells and rechargeable batteries to extend its potential lifetime to one year. Unfortunately, a command system failure terminated Courier after only seventeen days.

In addition to these early experiments with active repeater satellites, various organizations studied the efficacy of passive reflector satellites. Under Project West Ford, for example, the Air Force contracted with the Massachusetts Institute of Technology's Lincoln Laboratory to disperse 480 million copper dipoles, each 0.72-inch long and 0.0007-inch in diameter, in a nearly circular, nearly polar orbit in May 1963. This and other experiments led researchers to conclude that the passive systems were impractical when compared to repeaters.

Several active communications satellites

launched during 1962–65 revealed great technological strides. Two Telstar satellites, developed by Bell Telephone Laboratories for AT&T, established that multichannel telephone, telegraph, facsimile, and television signals could be transmitted across the Atlantic. The capacity of each Telstar included 600 one-way voice circuits or one television channel and 60 two-way voice circuits.

Two relay satellites, developed by the RCA Corporation for the National Aeronautics and Space Administration (NASA), had a more complex communications subsystem (two identical redundant repeaters) than Telstar did. These experimental systems demonstrated that existing technology could produce useful, medium-altitude communications satellites at a time when a more desirable geosynchronous altitude posed a somewhat daunting launch challenge.

Meanwhile, Hughes Aircraft Company supplied NASA with three Syncom satellites, which became the world's first geosynchronous communications platforms and, in the case of Syncom 3, the first geostationary satellite. The orbit- and attitude-control system developed by Harold Rosen and his team at Hughes Aircraft made simple, lightweight, geosynchronous satellites possible. This was, undoubtedly the single most important advancement in early SATCOM.

While the U.S. military has relied heavily on commercial SATCOM over the years, it also developed and launched dedicated systems to satisfy unique national security requirements. In 1958, ARPA had directed the Army and Air Force to plan for a strategic SATCOM system, with the Air Force responsible for the booster and spacecraft, and the Army for actual communications elements aboard the satellite as well as on the ground. Primary management responsibility for this geostationary system, dubbed Advent, resided with the Army. High costs, inadequate payload capacity, and an excessive satellite-to-booster weight ratio soon plagued this technologically ambitious undertaking, which someone described as a "not quite possible dream." To make matters worse, management problems obscured many technical issues and prevented a coherent resolution of them. Consequently, Secretary of Defense Robert McNamara canceled Advent in May 1962 and, pending a decision on whether the defense establishment could lease commercial satellite capacity to satisfy its requirements at lesser cost, delayed authorization of another dedicated military communications satellite program until July 1964.

The U.S. Air Force spearheaded the new effort, dubbed the Initial Defense Communications Satellite Program (IDCSP), and became responsible for procurement of all future military communications satellite systems. Intended for strategic communications, the IDCSP furnished the basic design for British Skynet and NATO satellites. Recognizing the advantages of a geostationary orbit for communications, the Air Force quickly evolved the sub-synchronous prototype IDCSP system of 1966 into the geostationary DSCS II of the 1970s and the jam-resistant DSCS III of the 1980s. The DSCS satellites basically provided service between large fixed terminals and transportable terminals with 20- and 8-foot diameter parabolic antennas.

Over time, another group of satellites

constituted the mobile-and-tactical segment of the MILSATCOM architecture. The first, Tacsat and its Lincoln Experimental Satellite, LES6, predecessor, were used experimentally to investigate various aspects of tactical communications on land, sea, and air. Developed by Hughes Aircraft Company, launched in February 1969, and operated by the Air Force Communications Service, Tacsat supported Apollo recovery operations by connecting aircraft with their carrier and ground stations. Because Tacsat failed in December 1972 and Fleet Satellite Communications (FLTSATCOM) capability was not expected before 1978, the Navy leased gap-filler UHF service from COMSAT General Corporation. Between 1978 and 1989, the Air Force procured from TRW Systems eight FLTSATCOM satellites and successfully launched six into geostationary orbits. In addition to sharing the FLTSATCOM satellites, the Air Force Satellite Communications system relied on packages aboard several satellites in high-inclination orbits to provide coverage of the northern polar region.

Based on congressional direction during 1976–77 to increase its use of leased commercial satellite services, DoD implemented the leased satellite program. It contracted with Hughes Communication Services in September 1978 for at least five years of service at each of four orbital locations. Five leased satellite launches occurred during 1984–90, and leases on three were extended into 1996.

The five subsystems that currently constitute the DoD MILSATCOM system, DSCS, Milstar, UHF, GBS, and the commercial adjuncts, will be discussed. We will look at how each of these subsystems evolved, describe their current configuration, and give examples of their impact on recent military operations.

DSCS

One can trace the roots of the DSCS back to cancellation of the U.S. Army's Advent program in May 1962. At that time, officials recommended two approaches: an IDCSP, which would use proven technology to develop simple satellites for placement, seven at a time, in random polar orbits at an altitude of approximately 5,000 miles using the proven Atlas-Agena launcher; and, somewhat later, an Advanced Defense Communications Satellite Program to place station-keeping satellites in synchronous orbits. A successful Titan IIIC launch in June 1965 led to that vehicle's selection for placing IDCSP satellites developed by Philco (later Ford Aerospace and Communications Corporation) into random, subsynchronous, equatorial orbits, three to eight at a time, between June 1966 and June 1968. Weighing only 100 pounds, these spin-stabilized satellites contained no moveable parts, lacked command and control capabilities, and had only a basic telemetry capability for monitoring purposes.

Under Project Compass Link in 1967, IDCSP provided pathways for transmission of high-resolution photographs between Saigon and Washington, D.C. As a result of this revolutionary development, analysts could conduct near-real-time battlefield intelligence from afar. By June 1968, IDCSP had been declared operational and its name changed to Initial Defense Satellite Communications System (IDSCS).

The follow-on DSCS II program aimed to overcome several deficiencies   limited channel

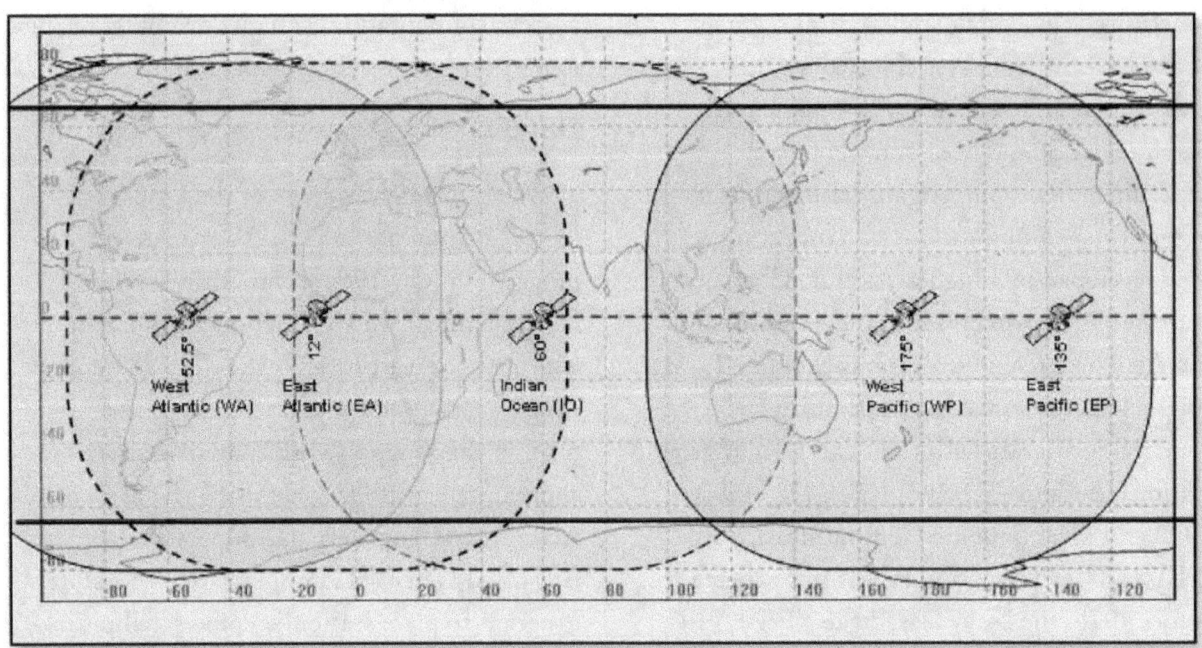

Figure 1. DSCS Space Segment.

capacity, user access, and coverage in IDSCS. In March 1969, TRW Systems received an Air Force contract to produce a qualification model and six flightworthy DSCS II satellites. Unlike the IDSCS satellites, DSCS II would have a command subsystem, attitude control and station-keeping capabilities, and multiple communication channels with multiple-access capability, and would occupy synchronous, equatorial orbits. It was dual-spin- stabilized and weighed approximately 1,300 pounds. Several design modifications extended the five-year design life to as much as twenty years. The Air Force launched the last DSCS II in 1989.

In 1974, the Air Force began design of an improved DSCS III satellite to meet the need for greater communications capacity, especially for mobile terminal users, and for better survivability. General Electric's DSCS III was three-axis-stabilized, weighed 2,475 pounds, and had a ten-year design life. The

first DSCS III launch to geostationary orbit occurred in October 1982. The Service Life Enhancement Program (SLEP) added high-power amplifiers to the last four DSCS III satellites to better support the warfighters. The last of these improved satellites was launched in 1989. One measure of the confidence the Air Force placed in the Jam Resistant Secure Communications capability afforded by DSCS III satellites is that, since December 1990, they have been the primary means for transmitting missile warning data from key sensor sites worldwide to correlation and command centers at Cheyenne Mountain, and elsewhere.

THE CURRENT DSCS

The DSCS space segment primary constellation consists of five DSCS III satellites in geostationary orbits as shown in Figure 1. Six residual satellites provide additional capacity for training, testing, and contingency operations.

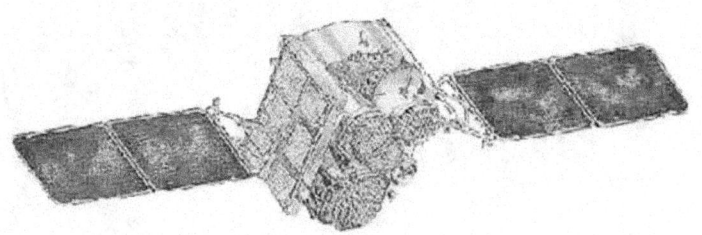

Figure 2. DSCS III Satellite.

The first DSCS III was launched in 1982. A total of fourteen have been launched, last one in 2003. A DSCS III satellite is shown in Figure 2.

The multiple-beam antennas (MBA) provide an important capability. The receive MBA has 61 beams. By suitably weighting these beams, the system can provide enhanced gain in the direction of friendly terminals and place nulls in the direction of interfering signals. The terminal segment includes:

- Fixed small, medium, and large terminals
- Transportable tactical terminals
- Mobile shipboard and airborne terminals
- Suitcase terminals

Example terminal types are shown in Figures 3 through 6.

The system operates in superhigh frequency (SHF) band in the vicinity of 8 GHz and uses 500 MHz of bandwidth. The major DSCS user communities include:

- Intelligence Organizations
- Combatant Commands (COCOMs) and their components
- Ground Mobile Forces
- Defense Information Systems Network (DISN)
- USAF Satellite Control Network
- Diplomatic Telecommunications Service

The DSCS system provides two types of capability. It provides high capacity in an unstressed environment (the absence of jamming and nuclear scintillation). In the current world situation, this is the dominant mode of operation. The system carries wideband traffic, high-speed computer-to-computer links, and interswitched trunks. The nominal DSCS satellite capability is about 75 Mbps between fixed sites, with an equivalent amount dedicated toward tactical sites. Capacity varies depending on the actual network requirements as changes occur in such factors as terminal performance, network topology, and satellite health.

The system can also operate in the presence of jamming by utilizing an antijam signal and the nulling capabilities of the MBA. In this mode, it provides secure voice and low-rate data services.

The basic DSCS III segment uses technology developed in the 1970s. However, the last four DSCS III satellites were upgraded as part of a Satellite Life Enhancement Program (SLEP). Technology upgrades included more transmit power in each channel, more sensitive receivers, improved solar cells to provide more power, and various processing changes. These upgrades improved the capacity by over 200 percent, improved support to small terminals, and provided greater flexibility in system usage.

Figure 3. Fixed Terminals DSCS AN/FSC–78 terminal (left) and AN/GSC–52 (right).

Figure 4. Transport Terminals - 20 foot AN/TSC–85 (left), 8 foot AN/TSC–93 (middle), and 2.4 meter triband (right).

Figure 5. Shipboard and Airborne Terminals —WSC–6 shipboard (left) and ASC–24 airborne on the E4B aircraft (right).

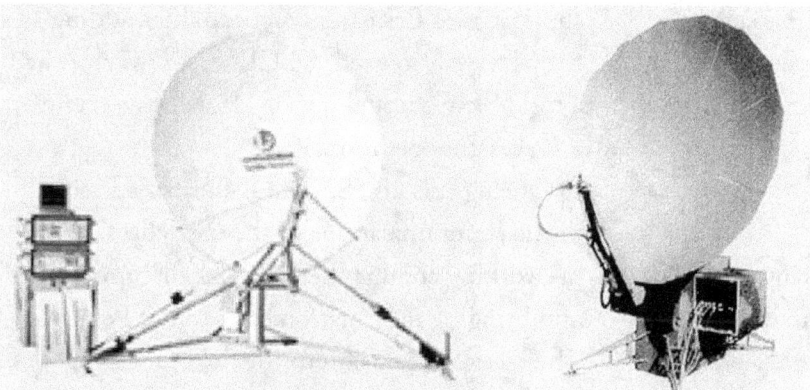

Figure 6. DSCS Suitcase terminals - USC–60A triband terminals (left) and LST–8000VT tri-band terminal (right).

The DSCS terminal segment has had continuing technology upgrades over the years. The most dramatic was the introduction of Light Multi-band Satellite Terminals. These terminals are packaged in transit cases and provide a rapid deployment capability that is important in support of current operational concepts. The first terminal of this type, the LST8000, was built by M/ACOM under USAF guidance and deployed in 1984 to support the White House Communications Agency. The devices have evolved into multiband terminals that provide SHF, C-band (46 GHz), and Ka-band (30/20 GHz) frequencies so they can operate over different satellite systems. The number of transit cases has been significantly reduced.

The requirements for wideband SATCOM have continued to grow rapidly. Warfighting requirements that contribute to this growth include:

- Situational awareness
  - imagery and intelligence, surveillance, and reconnaissance (ISR) products
  - environmental and map data
- Battle management
  - collaborative planning
  - video teleconferencing
  - simulation and wargaming
- Sensor-to-shooter
  - target sensing and tracking
  - target-weapon pairing
  - rapid target engagement
- Support
  - interactive data networks
  - telemedicine
  - reachback

Some specific examples of how DSCS has satisfied these warfighting requirements are provided in the next section.

OPERATIONAL IMPACT OF THE DSCS

Over the years, the DSCS has evolved into a highly successful system providing rapid, critical Command, Control, Communications, Computers, Intelligence, Surveillance, and Reconnaissance (C4ISR) communications for warfighters to execute military operations. Within hours, the system can provide high-capacity communications worldwide. The constellation is prepositioned and configured to handle large-scale operations. During peacetime operations, DSCS planners leverage the flexibility of DSCS and its ability to surge bandwidth toward quickly satisfying dynamic and robust user requirements. The historical support required for major contingencies has shaped the way MILSATCOM is managed today. Summaries of recent major conflicts that have changed the way DSCS is managed follow.

Operations DESERT SHIELD and DESERT STORM

The reaction to the invasion of Kuwait in 1990 triggered one of the most massive and rapid deployments of forces of that period. This rapid deployment produced a great dependence on MILSATCOM. The DSCS constellation serving the United States Central Command (USCENTCOM) area of responsibility (AOR) at the time consisted of a DSCS II over the Indian Ocean and a DSCS III over the Atlantic. Before the invasion, USCENTCOM employed four terminals in the region operating at aggregate of 4.5 Mbps. Within a month, forty-eight terminals were active at 38.3 Mbps, which equated to approximately 600 voice circuits. When President George H.W. Bush announced a 100,000-troop buildup, DSCS planners realized their overall capacity would be inadequate. As a result, the Joint Staff sanctioned the movement of another DSCS II from the western Pacific to the Indian Ocean region. In aggregate, the three satellites provided 68 Mbps of tactical communications using 110 SHF satellite Earth terminals. This provided critical Command, Control, Communications, and Intelligence (C3I) capability from USCENTCOM networks to the National Command Authorities via DISN services (secure voice, data, imagery, etc). The intelligence community also expanded its requirements from 23 Mbps to 36 Mbps. This put total DSCS support at 104 Mbps, a significant capacity during this timeframe. To augment DSCS

support, two British Skynet satellites adding 11 Mbps were also used. USCENTCOM thus had a total SHF throughput of 115 Mbps to conduct a successful operation.

During this era, SATCOM provided most of the communications to the battlefield, networking commanders with their components and bringing information back to the Pentagon. The tactical extension to the field primarily consisted of secure voice and data messaging services that were essential for communications. The typical SATCOM network consisted of intratheater mesh networks connecting tactical terminals with 256 Kbps data rates between sites. A few intratheater tactical trunks connecting the networks back to the DSCS Gateways provided DISN services to the field.

SHF capabilities supporting the Navy were very limited during this time. The Navy-mounted AN/WSC6 SHF terminal was limited to only a few ships, so the Navy obtained Air Force AN/TSC93B Ground Mobile Forces SHF SATCOM vans for aircraft carriers and amphibious flagships deployed to the Persian Gulf. This operation drove the need to accelerate installation of SHF capability onboard the large-deck ships for future operations.

Operation NOBLE ANVIL

During the 1999 crisis in Kosovo, requirements were developed quickly, and planning was intense. A clear need for communications had grown quickly since Operation Desert Storm. Within a very short timeframe, United States European Command (USEUCOM) had submitted requirements that quickly absorbed all available DSCS resources. Not only were the satellite resources a limiting factor, but the ground services were falling short as well. Because deployed tactical terminals were pulling DISN services from the Standard Tactical Entry Point sites, expectations were greatly exceeded. In the past, only a few links with small data rates were required to support an entire network on the battlefield. A major change was the increase in bandwidth required to support video applications. Video teleconferencing was used extensively during the conflict so that commanders could interface directly by video with their subordinate field commanders. Intelligence dissemination on the front line in support of the shooters also significantly increased video applications. The same 256 Kbps link used in Operation

Desert Storm now needed to be 2 Mbps or more. In response, the Defense Information Systems Agency (DISA) performed a rapid upgrade to all European Standard Tactical Entry Point sites. The final USEUCOM requirement totaled 64 Mbps, in addition to the 20 Mbps supporting USCENTCOM Operation Southern Watch and existing fixed requirements, which totaled approximately 100 Mbps. The same amount of throughput used in Operation Desert Storm that supported more than 120 deployed terminals now only satisfied the requirements for 20 deployed terminals. This was adequate for a small-scale operation, but it provided a perfect example of where SATCOM requirements were heading.

Wideband SATCOM Support

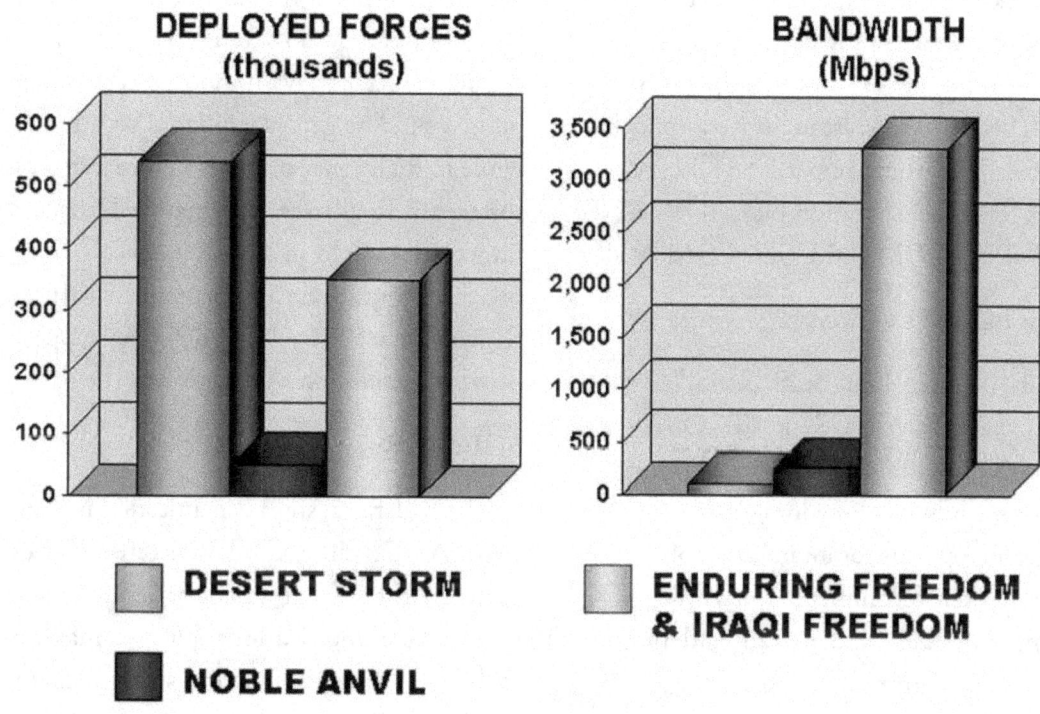

Figure 7. *Wideband SATCOM Increased significantly during the four major conflicts requiring less troop deployments.*

In addition, DSCS III satellites provided this support and were a huge improvement over the primarily DSCS II capability used during Desert Storm, since the DSCS III payload could be reconfigured more quickly and easily

Operation ENDURING FREEDOM

Hours after the attacks on the World Trade Center and Pentagon, DSCS support was critical for the immediate deployment of Navy ships along U.S. coastlines and en route to the Mediterranean. The Indian Ocean satellites were reconfigured immediately to provide added bandwidth and margin toward Afghanistan, with the expectation that deployment orders would soon follow. The satellite configurations were continuously updated as the operation developed. Over the coming months, tactical requirements increased to 120 Mbps in an area where existing coverage was already fully employed. DSCS provided the required bandwidth to support most of this operation, but the MILSATCOM resources were quickly running low. This would impact the follow-on operations for other operation plans (OPLANS) in the USCENTCOM and USEUCOM AOR.

Operation IRAQI FREEDOM

In April 2002, the Joint Staff assembled a team of experts to visit USCENTCOM to assist with potential conflicts. In addition to Operation Enduring Freedom, they were exploring options for an invasion of Iraq. Among their recommendations for overcoming shortfalls were upgrading deployed tactical terminals, improving capabilities at Standard Tactical Entry Point sites, relocating a DSCS satellite, and augmenting

DSCS satellites with commercial ones. DSCS support of warfighter requirements surged to 350 Mbps tactical and 150 Mbps fixed. These requirements were supported with one DSCS SLEP and three DSCS III satellites. The total capability provided for this operation far exceeded anything to date and set a precedence for the way OPLANS are developed and communications are provided.

Requirements for SHF Earth terminals were unprecedented. In an era when military planners were talking about megabits per second, gigabits per second were now required. With the introduction of unmanned aerial vehicles (UAVs) and huge amounts of digital imagery and data, large communications pipes were required to get the information to the shooters in real time. During Operation Desert Storm, air-support operations were required to have target data available before takeoff. The Iraqi military would often move targets before the actual attack. With modern technology, pilots would be getting updated coordinates while in flight, just minutes before impact on the target. Real-time information feeds and targeting resulted in a greater success rate of precision munitions and ordinance hitting their targets. SHF MILSATCOM support provided data such as this to win the battle.

MILSTAR

The Milstar system was initiated in April 1981. At that time, Cold War tensions were high and the system was designed to ensure essential communications during a major nuclear exchange. It was designed to survive an anti-satellite attack using nuclear weapons, counter the projected Soviet jamming threat, and operate autonomously for an extended period. The consequence of these requirements was a Milstar system that provided only low data rate (LDR) service. However, it is important to realize that Milstar guaranteed the necessary service in any anticipated environment.

President Ronald Reagan assigned Milstar "Highest National Priority" status in 1983, which allowed the program to proceed with few funding restrictions and led to the addition of numerous technical requirements to meet more varied missions.

Milstar marked a major change in the role of satellites. All of the communications satellites prior to Milstar acted as relays in space. The Milstar constellation, with its crosslinks and onboard signal processing, provided a network in space that allowed communications around the globe without intermediate ground stations.

Initially designed to provide LDR-enhanced extremely high-frequency (EHF) communications, Milstar offered crosslink capabilities and extensive hardening against radiation. Four satellites would be placed in polar orbits, and four others, in geostationary orbits. Because the primary objective was survivability, not high capacity, the first two satellites (Milstar I) carried payloads capable of transmitting voice and data at the LDR rate of 752,400 bps. Each satellite supports 192 channels of service.

The end of the Cold War prompted a Milstar restructure referred to as Milstar II. The revised system provides medium data rates (MDR) of 4,800 bps to 1.544 Mbps. Additionally, the program underwent significant downsizing based on congressional demands and DoD reviews.

By early 1994, the Milstar program included only six satellites without the vast array of survivability features and with fewer ground control stations. The first block of two satellites, designated Milstar I and built by Lockheed Martin, would retain the limited-use LDR capability, but the subsequent Milstar II satellites would be equipped with an MDR package to better support tactical users. On 7 February 1994, the first Milstar satellite went into orbit. Even before it had completed its on-orbit checkout, the 10,000-pound satellite contributed operationally to Uphold Democracy, the U.S. intervention in Haiti. Each Milstar satellite serves as a space-based "smart switchboard" by actually processing communications signals, crosslinking with other Milstar satellites, and directing traffic from terminal to terminal anywhere on Earth. The need for intermediate ground relays under normal conditions is thus virtually eliminated. The last of six Milstar satellites went into geostationary orbit in April 2003.

Current Milstar System

The Milstar system operates in the EHF/SHF band (44 GHz uplinks and 20 GHz downlinks) and utilizes 2 GHz of bandwidth on the uplink and 1 GHz of bandwidth on the downlink. The EHF band allows the use of narrow beams, so less transmitted power and smaller antennas are allowed. Jammers must be physically closer to be in the satellite's beam. In a nuclear environment, outage times are much lower at EHF. However, EHF signals are affected by rain and foliage, so large link margins are required.

Important Milstar system features include:
- Robust signal waveform to provide antijamming and nuclear protection

- Onboard signal processing (a switchboard in space)
- Intersatellite crosslinks at 60 GHz
- Cross-banding between EHF/SHF and UHF
- Flexible networking
- A sophisticated antenna farm

The Milstar II satellite shown in Figure 8 has an LDR wing and an MDR wing. The spacecraft is 78 feet long, 116 feet wide, and weighs 10,100 pounds. The picture shows the extensive antenna farm, one of Milstar's most impressive features. The transition from Telstar to Milstar is an outstanding technological accomplishment.

The terminal segment includes:
- The Air Force command-post terminals which support LDR and are nuclear-hardened
- The Navy terminals which support both LDR and MDR
- The Army Secure Mobile Antijam Reliable Tactical Terminal (SMART-T) which supports LDR and MDR
- The Single Channel Antijam Man-Portable (SCAMP) II terminal which supports LDR

Because the EHF frequency is significantly higher than the SHF, the antennas can be much smaller. A picture of a SMART-T terminal is

Figure 8. Milstar II Satellite.

Figure 9. Milstar SMART-T terminal.

Figure 10. Milstar SCAMP II terminal.

shown in Figure 9, and a picture of the SCAMP II terminal is shown in Figure 10.

Milstar serves strategic-level, theater-level, and tactical-level users. At the strategic level the system provides tactical warning/attack assessment data relay, nuclear command and control (C^2) conferencing, emergency action message dissemination, and force managements and reportback. At the theater/tactical level it provides C^2 communications for joint task forces and Army Corps, and below, units, tactical intelligence dissemination, range extension for the Army mobile subscriber equipment, dissemination of ATOs, Tomahawk cruise missile updates, and Navy task force connectivity.

Operational Impact of Milstar

Milstar, as a communications system, has the flexibility to be reconfigured to meet changing operational requirements. The system supports a wide range of strategic and tactical missions, including:

- Connectivity for C^2 of tactical forces.
- Connectivity for deployed Special Operations Forces.
- Connectivity for deployed naval battle groups to support rapid deployments of land, air, and naval forces anywhere in the world.
- Missile threat conferences.
- Nuclear force execution orders.
- Air Expeditionary Force en route planning
- Reportback information from strategic and nonstrategic nuclear forces.
- C^2 connectivity between COCOMs and their components.

Because of its inherent design features, Milstar can support these missions despite enemy jamming and can, if necessary, withstand nuclear effects.

Operation DESERT STORM

EHF only provided experimental service during Operation Desert Storm using the EHF package on the UHF Follow-On (UFO). Although EHF terminals were just being fielded, the EHF package on the Navy's FLTSATCOM satellite was used to provide a secure communications link between the Chairman of the Joint Chiefs of Staff and the Commander of USCENTCOM. Contingency plans existed to bring transportable EHF terminals to the theater if our MILSATCOM systems were jammed.

Operation NOBLE ANVIL

By the time of Noble Anvil, enough EHF terminals had been delivered to the military Services so that EHF was integrated into the warfighters' communications plans. Milstar support to operations in Kosovo began in June 1999 and continues today. In particular, the Navy had installed EHF LDR terminals in the Battle Groups. In general, LDR was used for Naval tactical communications, including ATO dissemination. Figure 13 depicts transmission times for ATOs and other information exchanges at different data rates. Communications support was also provided to deployed Kosovo ground forces.

Operations ENDURING FREEDOM and IRAQI FREEDOM

Now that the MDR constellation was nearly complete and substantial numbers of EHF terminals were deployed, EHF saw more extensive use during these conflicts. The Navy and Marine Corps both used MDR extensively for a variety of applications.

The Milstar system has been critical for secure communications, Global Hawk surveillance, and "net-centric" operations. Critical U.S. intelligence, such as National Reconnaissance Office and Central Intelligence Agency data, is similarly routed from the United States to the theater. The system's space-to-space relay capability avoids the need to bounce this secret communications traffic through multiple ground stations.

In addition to operations involving Global Hawk surveillance and Special Forces missions, the Marine Expeditionary Forces in Iraq have been exploiting Milstar's secure communications, antijamming capabilities for tactical ground operations. The Army's 4th Infantry Division, moving northward from southern Iraq, also is heavily equipped with Milstar terminals, sources said. According to the commander of 124th Signal Battalion,

> The SMART-T is an incredible force multiplier and it has performed remarkably well. My CG's [Commanding General's] Assault Command Post (ACP) had a Battle Command on the Move Bradley equipped with ABCS [Army Battle Command System] systems. We put Line of Sight (LOS) and SMART-T teams with him as he deployed into Iraq. His SMART-T crew was able to establish voice and data communications whenever he decided to halt for over 30 minutes. They averaged 15 minutes to get the link in and were able to do it as fast as 9 minutes.

> We put the same packages with the 1st Battalion, 44th Air Defense Artillery (ADA) Tactical Operation Center (TOC), 1st Brigade Combat Team (BCT) Commander, 1st Squadron, 10th Cavalry TOC and had similar results. At one point we had a BCT in Iraq; a BCT doing Reception, Staging, Onward movement, and Integration (RSOI) in Camp New York; and a BCT at the port off-loading equipment. We would not have been able to accomplish the mission without the SMART-T.

The 124th Signal Battalion (assisted by other units) was able to install, operate, and maintain the largest division communications network in the history of the Army consisting of eleven Node Centers, more than forty-five Small Extension Nodes, four Node Center Support Elements, and fourteen SMART-Ts dispersed over a 90,000 square kilometer area of operations in Iraq. Again, it would have been impossible to do this mission without the SMART-Ts.

Milstar Technologies

The Milstar system was designed to emphasize robustness and flexibility, which resulted in many significant technological enhancements. Robustness is the ability to operate under adverse conditions, including direct jamming, interception, and nuclear attack; flexibility is the ability to provide worldwide, unscheduled access and worldwide connectivity to terminals on all types of platforms. Satellite features to support system robustness include frequency-hopped uplinks and downlinks, extensive onboard processing, 60 GHz crosslinks, and nuclear hardening. Satellite features to support flexibility include multiple uplink and downlink channels at various rates, in-band control channels for service requests, multiple uplink and downlink beams (including agile beams that can switch on a hop-by-hop basis), and routing of individual signals between uplinks, downlinks, and crosslinks. To achieve robustness and flexibility, Milstar 1 technology developments included broadband multichannel demodulators, hardened general-purpose processors, advanced piece-parts, and crosslink technologies.

Access control messages are processed by onboard processors, which have been hardened using special piece-parts and shielding. Access control responses and other control data are generated by the onboard processors and sent back to the user terminals on the downlink. Through access control messages, Milstar users may request various communication services such as joining an existing network, establishing a new network, making point-to-point calls, moving a spot beam antenna, and requesting information about the status of the Milstar constellation.

The Milstar Spacecraft Processor was a general-purpose, radiation-hardened Mil 1750A computer developed specifically for Milstar I. One computer was used to control the spacecraft bus; another, to control the payload. The Milstar Spacecraft Processor uses a 2-micron Complimentary Metal-Oxide Semiconductor (CMOS) radiation-insensitive digital processor chip set developed and manufactured for Milstar by the Sandia Corporation, and radiation-insensitive volatile Static Random Access Memory memory manufactured by Harris

Corporation, using Sandia-licensed designs. All of the Milstar satellites nonvolatile digital memory employs magnetic bubble-memory technology. The demodulation processing implemented for Milstar I used an acousto-electric array of processor technology that at the time was the only known way to demodulate the approximately 200 Frequency Shift Keying channels in a package with sufficiently low weight and power requirements to be applicable to spaceborne operations. The antenna suite includes an agile antenna, which is capable of switching between coverage areas across the field of view in nanosecond timescales. This required development of fast ferrite switches and a radio-frequency (RF) lens structure. The satellite design included crosslinks at 60 GHz, which required the development of 60-GHz sources, receivers, and antenna manufacturing technologies to allow the construction of large reflectors with the required surface smoothness and suitability for space operation. Any one of these technologies would have been considered a major advancement; Milstar I tackled them all.

The MDR payload has eight narrow spot-beam antennas designed to meet Army and Navy requirements. Two of the MDR spot-beam antennas have onboard adaptive nulling capability to negate the effects of both in-beam and out-of-beam jammers. The six other small spot-beam antennas without nulling are called Distributed User Coverage Antennas. The onboard autonomous nulling antenna design was a significant technology development for the Milstar II program. The nulling antenna is a complete feedback-control system designed to continuously maximize desired signals while processing-out jamming signals. Nulling antenna technology combined with spread-spectrum processing can provide antijam protection against both in-beam and out-of-beam jammers, even when the desired user is operating a low-power terminal at a relatively high data rate.

While the Milstar LDR and MDR payloads share many architectural features, the MDR payload developed in 1991–92 incorporates many of the technological advances made since the mid-1980s. Technology advances occurred in both digital and microwave integrated circuits. These new integrated circuits are key to the implementation of the MDR payload within the weight and power constraints of Milstar II. The LDR payload digital processing subsystem is based on 1.5-micron CMOS custom large-scale integrated circuits. The maximum number of digital gates per device is approximately 5,000. The LDR processor chip sets, excluding the primary onboard computer, contain 35 custom large-scale circuit designs, which are reused in multiple processor applications. Each Milstar LDR payload has 630 of these custom integrated circuits. The processors in the MDR digital processing subsystem are based on radiation-hard 0.8-micron CMOS application-specific integrated circuits. The 0.8-micron CMOS application-specific circuits can accommodate up to 100,000 gates per device. The MDR digital subsystem, excluding the primary onboard processor, has only fourteen unique application-specific integrated circuit designs. The total number of large-scale integrated circuit devices required for the MDR processors is 397.

The MDR RF equipment took advantage of advances in Gallium Arsenide (GaAs) Monolithic Microwave Integrated Circuit technology. As an example, the nulling antenna low-noise amplifiers are fully integrated, four-stage High Electron Mobility Transistor amplifiers, all on a single chip. These advances in piece-parts and circuit design result in significant weight and power savings. The LDR payload equipment totals approximately 2,400 pounds and 1,500 watts, while the MDR payload equipment is approximately 1,000 pounds and 1,000 watts. Much of the reduced weight and power of MDR is due to the use of advanced RF and digital technologies.

UHF SYSTEM

Evolution of UHF Communications

The early history of UHF SATCOM was described above. By 1988, the UHF system had evolved from the early mixture of FLTSATCOM and leased satellites into a fourth-generation capability called the UFO system.

This UFO system continues to operate today and consists of eight satellites plus an on-orbit spare. It was designed to replace the Navy's aging FLTSATCOM and leased satellites as well as accommodate a national growing requirement for UHF capacity. In July 1988, the Navy awarded Hughes Space and Communications Company (now Boeing Satellite Systems, Inc.) an innovative, fixed-price acquisition contract giving it the latitude to select commercial-off-the-shelf components for the UFO satellite and procure commercial launch vehicles for putting UFO capability on orbit. Following an initial launch failure, Boeing Satellite Systems successfully launched the next nine UFO satellites between 1993 and 1999. The first three satellites supported mobile communications and fleet broadcast services with UHF and SHF payloads.

The next four satellites carried an additional EHF capability to provide protected communications support using Milstar terminals.

Current UHF System

The UHF system operates in the 225400 MHz frequency band, coexisting with line-of-sight UHF requirements. It is the primary system employed for mobile SATCOM user communications. The associated small and inexpensive terminals routinely use nondirectional antennas. The available UHF bandwidth is limited by both the frequency and propagation characteristics of the system. User demand exceeds the available capacity. UHF communications are fundamentally unobstructed by weather or foliage, but unfortunately the UHF band provides virtually no antijamming capability. Communications can be degraded by changing propagation characteristics dominated by ionospheric scintillation, multipath, or other unintentional user interference sources.

The current UHF space segment consists of eight UFO satellites in six inclined geosynchronous orbits. Two satellites are located in each of four coverage areas. The system provides 39 channels on each satellite, 17 of which are 25 KHz, and the remaining 21, 5 KHz. An additional service is provided by a fleet broadcast channel consisting of a jam-resistant SHF/EHF uplink and associated UHF downlink. All UHF channels operate as "bent pipes" through a simple, transponded satellite. The most recent four enhanced UHF satellites provide an EHF payload that operates with Milstar terminals as well as a Ka-band (20 GHz) broadcast package that supports the GBS

Figure 11. Army UHF Spitfire.

(described later).

The Army, Navy, and Air Force all employ UHF terminals to fulfill a portion of their communications needs. Efficient system use is achieved through implementing demand-assigned multiple-access techniques to allow resource sharing. The development of a workable demand-assigned, multiple-access system was an important technical accomplishment. A representative terminal, the Army PSC5, or Spitfire terminal, is shown in Figure 11.

Typically, Army users include early-entry forces and a variety of mobile communication users. Air Force airborne and manpack tactical communications, Airborne Warning and Control System, and Take Charge and Move Out operations also take place over this UHF system. The Navy's Fleet Satellite Broadcast subsytem, secure voice networks, and information exchange systems operate on UHF as well.

The key technical challenge in the UHF system continues to be improving the effectiveness of the demand-assigned, multiple-

195

access implementation to increase capacity even further.

Operational Impact of the UHF System

The operational history of UHF spans many years. The numerous uses of this critical system stem directly from the advantageous propagation characteristics of communicating at these frequencies. While frequencies higher than UHF perform as effectively in clear weather when the view of the distant end is unobstructed, reduced capability can be expected when rain, jungle foliage, or other impediments that cause signal degradation are present. UHF, on the contrary, is more robust and likely to be less expensive under these types of conditions. Such communication links routinely operate with higher availability and reliability and under a more varied set of conditions. It is this characteristic of UHF that makes it particularly well suited to tactical communications as well as to other specialized missions. UHF satellites provide military planners with a reliable transmission medium for sending critical intelligence, operations, and logistics data.

UHF communications over the last decade have played an important role in all our military operations. This medium has been critical in every conflict: Desert Shield, Desert Storm, Bosnia, Kosovo, Afghanistan, as well as the push to Baghdad. It will continue to be of vital use as our forces work nation-rebuilding issues in Iraq.

UHF SATCOM routinely allows commanders to overcome many of the distance and terrain restrictions that face similar broadcast radio networks operating at higher frequencies. At the tactical level, this enables units to operate informal voice nets over wide areas without deploying VHF FM rebroadcast stations. The small portable terminals provide maneuver commanders the ability to maintain control over subunits under adverse circumstances.

In both Bosnia and Kosovo, battlefield commanders routinely used UAVs operating over secure UHF SATCOM paths to assist in assessing the local operational situation. Deployed worldwide, these systems support joint combatant forces in peacetime, wartime, and antiterrorism operations. High-mobility vehicles and their accompanying ground-based system components transport information and imagery to assist in identifying targets and in other intelligence-gathering activities.

In Operations Desert Shield and Desert Storm, these UHF capabilities were used to provide strategic, operational, and tactical communications support to U.S. Navy battle groups. In particular, it facilitates command and control of guided missiles, like Tomahawks and other cruise missiles, from ships and submarines. It provides a path for updating cruise-missile missions from bases ashore. UHF capability is also used in disseminating ATOs to B52 bombers and providing a medium for effective mobile-user communications for national and coalition forces.

A very heavy demand and subsequent dependence on these communications has developed more recently as a result of our efforts in Afghanistan. To provide for the additional required capacity, the U.S. military reassigned a nine-year-old Navy spacecraft, the UFO F2, to the USCENTCOM theater of operations to assist in satisfying the increased demand for vital battlefield communications. In addition to combat missions, UHF radios are routinely used in communications for air traffic control coordination and in secure communications via the Secure Telephone Unit III, and tactical commanders can send Secret Internet Protocol Router Network e-mail over UHF communications paths for entry into large terrestrial networks.

On 7 April 2003, during Operation Iraqi Freedom, the United States executed an air strike against the Iraqi leader Saddam Hussein. Tipped off by someone on the ground, the information regarding Saddam Hussein's location was relayed through a UFO satellite to a B–1B bomber already in the air. The B–1B hit the target with four Global Positioning System guided Joint Direct Attack Munitions. This demonstration of flexible response was made possible because of the UFO system.

Evolution of GBS

One can trace the origins of GBS back to an April 1992 final report to Congress on the conduct of the Persian Gulf War. That report highlighted the limited capability of existing military and civilian SATCOM systems to provide responsive, high-capacity links of the kind needed for imagery and video transmission to deployed, mobile tactical users with small antennas. In early 1993, the National Information Display Laboratory, hosted by Sarnoff Corporation, began briefing government officials on the value of direct broadcast satellite (DBS) technology. An industry team led by Hughes Aircraft Company had been developing this technology for several years with the intention of marketing a new

commercial broadcast service Direct Television. Through a series of detailed briefings, laboratory personnel convinced key DoD decision makers that this technology had significant potential for overcoming the communications shortfall identified in the Persian Gulf War report.

The next steps toward GBS involved a series of demonstrations. In a project designated Radiant Storm, the National Information Display Laboratory helped the Navy accomplish the first transmission of encrypted intelligence data via a DBS system. Next, with the laboratory's support, the Air Force undertook a more ambitious program that yielded several key results:

- Use of commercial Ku-band satellites to achieve worldwide coverage
- Use of commercial, mobile uplinks for theater injection
- Linking to a UAV ground station in-theater for live video dissemination
- Development of a high-speed data interface unit for the commercial encoder
- Development of asynchronous transfer modecompatible interfaces at both the encoder and receiving stations
- Use of standard telephone links combined with DBS service to provide two-way, interactive capabilities

Successful application of these results during the 1995 Joint Warrior Interoperability Demonstration was termed as one of the demonstration's "golden nuggets." Based on this experience, the National Information Display Laboratory proposed using this same DBS technology to disseminate UAV video in support of Bosnian operations during the summer of 1995.

Current GBS

The GBS is a combined space and C3I system that provides one-way, high-volume information flow (data files, imagery, and voice) to garrisoned, deployed, or on-the-move forces. It is the successor to the Joint Broadcast Service (JBS) that played an important role in the Bosnian operations. It consists of three subsystems:

- A broadcast management and signal injection subsystem
- A space segment which broadcasts information to the users
- User terminals

The signal injection subsystem consists of a primary injection point in Norfolk, Virginia, which can uplink 94 Mbps to the space segment, and theater injection points which can uplink 6 Mbps to the space segment. The uplink is at Ka-band (30 GHz).

In March 1996, the Navy ordered a high-power, high-speed, Ka-band GBS payload to be added to UFO satellites 8, 9, and 10. Derived from Hughes's experience with commercial Ku-band satellite broadcast systems in Bosnian operations and elsewhere, the GBS package supplied data delivery rates vastly superior to any prior MILSATCOM capability. The first GBS payload went into service aboard UFO Flight 8 in June 1998, and the launch of UFO Flight 10 in late November 1999 completed a three-satellite constellation providing DoD near-global broadcast coverage. Transmitting to small, mobile, tactical terminals, the GBS package revolutionizes the full range of DoD's high-capacity communications requirements, which ranged from intelligence

dissemination to quality-of-life programming.

The current space segment consists of Ka-band payloads on the last three UFO satellites. However, because only a few UFO satellites hosted the Ka-band GBS payload, it became necessary to continue leasing Ku-band commercial satellite services to augment UFO where gaps in coverage existed and, if necessary, to complement the limited number and size of the downlink beams from UFO GBS.

The payload has two uplink antennas: one is fixed and the other is steerable. It has three steerable downlink antennas: two are narrow-beam (500 nm) and one is wide-beam (2,000 nm). Four transmitters operate in the 20-GHz range. The use of narrow beams allows high transmission rates (23 Mbps) to reasonably small terminals.

The user terminal subsystem consists of land-based fixed and transportable terminals with 1-meter antennas and ship-based terminals.

The entire system is depicted as a block diagram in Figure 12.

The GBS concept of operations includes two types of information management referred to as "smart push" and "user pull." Smart push information is that which can be predetermined on the basis of a user's mission. User pull information is that requested by a user via a low-capacity feedback link and is broadcast at a high data rate.

Typical information that can be provided over GBS includes:

- Integrated Broadcast Services
- National Imagery (National Imagery and Mapping Agency)
- Theater Data (COCOMs)
- ATO (Joint Forces Air Component

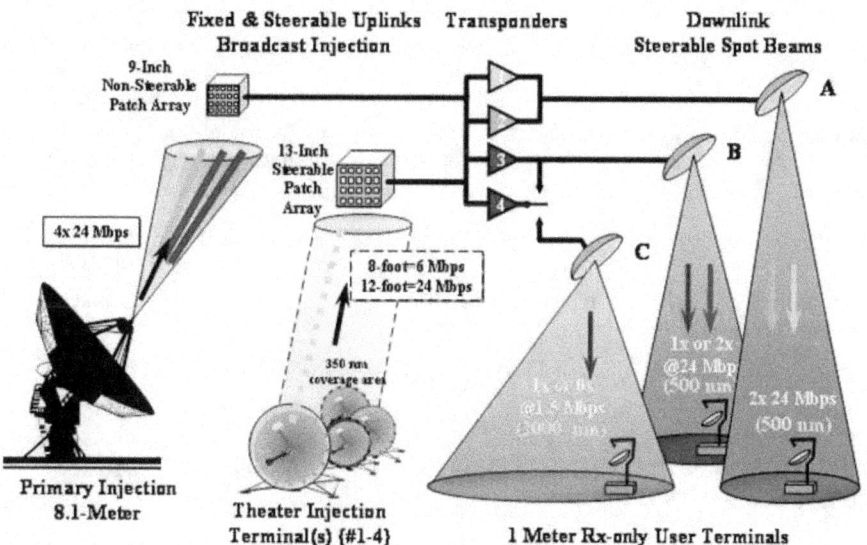

Figure 12. GBS System.

Command)
- Fleet Broadcast (Navy)
- UAV Target Video (airborne link segment)
- Joint Deployable Intelligence Support System products (multisource)
- Weather Reporting (multisource)
- Tomahawk Mission Data (Navy)
- Situation Awareness (services)
- Logistics (services)
- Training (services)
- Army Forces Radio and Television System
- Television Receive-Only, Cable News Network, Movies

To illustrate the advantage of the GBS 23 Mbps data rate, Figure 13 shows the transmission time for representative products. One of the key warfighting advantages of GBS is that it allows near-real-time targeting based on sensor inputs (e.g., Predator video). In a typical application, an encrypted Predator video is downlinked to a terminal, injected into the GBS payload, and broadcast to the appropriate users.

Operational Impact of GBS

The GBS concept provides the correct information to the user at the required time and place. The fundamental approach addresses the standard push-and-pull GBS architecture capable of multiple security levels, and tailors the information to a specific user's needs by sensor-to-shooter couplings. The GBS has met with resounding success over the last ten or so years in revolutionizing military use of information. This is reflected in an increased understanding of battlefield events and an improved ability to apply destructive force when and where it is necessary. GBS expands the use of battlespace dominance through improved information use. In the pre-GBS environment, serious bandwidth problems arose with the use of UHF circuits during Operation Desert Storm when transmission of a 500-page ATO and its associated imagery could take hours. More recently, with the use of direct broadcast

SATCOM Throughput Example Information	2.4 Kbps (for example, on) Milstar & UFO	64 Kbps (for example, to) Navy's IT-21	512 Kbps (for example, to) SIPRNet	1.544 Mbps (for example, on) Milstar MDR	23 Mbps GBS
Air Tasking Order (DESERT STORM) 1.1Mb	1.02 hr	2.61 min	17.19 sec	5.7 sec	.38 sec
Tomahawk Mission Data Update 0.03 MB	100 sec	4.29 sec	.47 sec	.16 sec	.01 sec
Imagery 8x10 Annotated 25 Mb	22.2 hr	57 min	6.25 min	2.07 min	8.4 sec
Desert Shield Time Deployment Data (log support) 250 Mb	9.65 day	9.92 hr	1.09 hr	21.59 min	1.45 min

Figure 13. Transmission Time for Representative Products.

transponders, the same data is transmitted within seconds. This capability makes it possible for combat leaders to be keenly aware of the situation confronting them. The information they receive helps them assess the enemy's size, location, and activity. This instant situational awareness now available to troops and pilots is achieved by integrating satellite intelligence, UAV flights, and ground-signals intelligence stations into a common picture.

Operation NOBLE ANVIL

Originally implemented as the Bosnia JBS, the first phase of the GBS became operational to support immediate operations in Bosnia. It provided a tactical network allowing virtually instantaneous communications among all U.S. forces. In large part, this technology provided for dissemination of UAV video in support of Bosnia operations during the summer of 1995. This represented a significant improvement over the Gulf War situation when only a limited use of in-theater video, and then, only to commanders at high levels, was available. In Bosnia, surveillance and reconnaissance assets plus live video feeds from intelligence assisted in conducting operations and gathering intelligence. The video feeds were provided at the division level. The trend to push information to lower command levels continues as battlefield video-teleconferencing becomes available down to brigade and now, in some cases, battalion level. UAVs are proliferating in the battlespace. Predator UAVs that became operational in Bosnia around 1995 used this system and flew more than 600 missions in support of NATO, UN, and U.S. operations. JBS commercial SATCOM service used commercial television DBS technology modified for military functions. JBS was implemented as an element of the Predator UAV communications architecture specifically for the dissemination of electro-optical and infrared video sensor information. The processed video information was relayed from the UAV ground station via transoceanic cable to the JBS injection site at the Naval Research Laboratory in Washington, D.C., where it was retransmitted to the Atlantic Ocean Region and then JBS receivers. The system was called the Bosnia Command & Control Augmentation (BC²A) Program's JBS.

Operations ENDURING FREEDOM and IRAQI FREEDOM

The GBS system now "pushes" weather and a variety of data, imagery, other high-volume intelligence, as well as other information, to a widely dispersed user community through relatively low-cost receive terminals. The system includes the capability for users to request, or "pull," specific pieces of information. The GBS distributes many high-bandwidth products directly from the United States to the lowest levels of command. GBS traffic runs the gamut from video and large data files to Internet Protocol traffic. GBS collects real-time intelligence from ground-based and space-based sources, collates that intelligence, and sends it immediately to fighter jet cockpits and mobile Army support vehicles. This connectivity provides more than enhanced situational awareness. It has changed the roles of the warfighter and the weapons platform. Predator follows a conventional launch sequence from a semiprepared surface under direct line-of-sight control. Takeoff and landing typically require 2,000 feet. Mission control is achieved through Ku-band satellite links or line-of-sight data links to produce continuous video. Video signals received at the Ground Control Station are passed to the Trojan Spirit van for worldwide intelligence distribution or directly to operational users via a commercial GBS. Command users can task the payload operator in real-time for still images or real-time video. The local commander has excellent visibility of his battlespace, enhanced largely by the communications capability of GBS.

Critical GBS mission-traffic continues to support operations in both Afghanistan and Iraq. In support of Operation Iraqi Freedom, available bandwidth has been doubled throughout the duration of the conflict. The future of GBS likely includes netted Joint Service Command, Control, Communications, Computers, and Intelligence (C⁴I) systems operating over the GBS.

201

COMMERCIAL ADJUNCTS TO MILSATCOM

Over the past forty years, numerous U.S. military organizations have relied on commercial SATCOM to meet specific, ad hoc requirements or to supplement dedicated military capabilities. Officials have reasoned that leased commercial channels provide cost-effective voice and data communications and meet the growing demand for long-haul, wideband services such as computer-to-computer nets, video conferencing, high-speed facsimile, and electronic document transfer. Despite the obvious advantages of tapping into commercial SATCOM capabilities, the ongoing challenge remains to make the most efficient use of them. That depends, in turn, on effectively integrating them with MILSATCOM capabilities to form a relatively seamless architecture.

Next, we describe several commercial adjuncts and their operational usage. Before we begin our general discussion however, it is important to understand that during the most recent periods of conflict—Operation Enduring Freedom and Operation Iraqi Freedom—DoD for the first time put commercial bandwidth in place ahead of the warfighter and weapons systems relying on it. In doing so, DISA took risk in anticipation of requirements by using a combination of contract vehicles and vendors to develop an architecture flexible and agile in its design that ultimately satisfied the myriad of requirements asked for by USCENTCOM and its components.

Commercial Adjuncts

Three types of commercial satellite service are used to augment the military satellite systems.

The first is the fixed satellite service provided by INTELSAT and other carriers. The space segment consists of geostationary satellites operating at C-band (6 and 4 GHz) and Ku-band (12 and 14 GHz). The service provided is similar to the DSCS system operating in an unstressed mode. The system provides high throughput for imagery and video and gives a surge capability during crises and wars. A technical development enhancing the utility of the system is the tri-band (C, SHF, and Ku) transportable transit-case terminal. This provides the user with a terminal that can be used with either the DSCS or INTELSAT system.

The second type is the mobile satellite service that uses geostationary satellites. The International Maritime Satellite (INMARSAT) system operates at L-band (1.5 GHz) and provides voice and video service to very small terminals. The system is widely used by reporters accompanying military operations. It is also used to provide quality-of-life communications for deployed forces.

The third type is the mobile satellite service that uses satellites in low Earth orbit. The smaller range to the satellite allows the use of handheld terminals only somewhat larger than a cell phone.

Operational Impact: Desert Shield and Desert Storm

During Operations Desert Shield and Desert Storm, Commercial T1 circuits were leased over two INTELSAT satellites. Fourteen commercial satellite terminals were used providing 3 Mbps of data (simplex). This equated to 25 percent of the SATCOM used in the theater. Additionally, INMARSAT was used to supplement UHF SATCOM. For the first time, MILSATCOM (DSCS, allied, and commercial) was the bread-and-butter source of communications connectivity, both long-haul and tactical.

One lesson learned was that DoD should procure through the military departments commercially available satellite terminals and bandwidth. On-call arrangements for commercial transponders should also exist. In 1994, DISA was funded to develop and implement a commercial SATCOM program, then known as the Commercial SATCOM Communications Initiative. On 1 October 1998, the initiative was transitioned to the Defense Working Capital Fund. This transition made customers responsible to pay for the bandwidth leased to support their requirements. Without central funding, DoD was unable (through DISA) to preposition commercial satellite bandwidth and to pay for on-call arrangements for additional bandwidth. The result was that during Operations Noble Anvil, Enduring Freedom, and Iraqi Freedom DISA again was forced to lease bandwidth as funding was made available. Fortunately, sufficient bandwidth was found to support all three operations.

Operational Impact: NOBLE ANVIL

In Bosnia and Kosovo, the use of commercial augmentation to MILSATCOM provided critical extension of DISN services (Secret Internet Protocol Router Network, Nonclassified Internet Protocol Router Network, Defense Red Switch Network, Defense Switched Network, video teleconferencing, etc). This was the first time that commercial SATCOM was the dominant provider of SATCOM. Transponders leased during this conflict supported DISN extension into the Balkans; UAV control and video; C2 networks, Kosovo forces intratheater communications; situational awareness tools (Blue Force Tracking); and Joint Task Force requirements. In January 1996, USEUCOM requested an extension of the DISN into the Balkans to help free up tactical assets. The DISA-Europe field office installed DISN points-of-presence consisting of Integrated Digital Network Exchange multiplexers and routers in Bosnia, Croatia, and Hungary. These points-of-presence provide long-haul connectivity back to three locations in the central region of Germany. During Operation Allied Force, USEUCOM again requested DISA to extend the DISN into Kosovo and Macedonia. DISA now had points-of-presence in Skopje, Macedonia, and Pristina, Kosovo. USEUCOM has used over forty-four European leased circuits at 2.048 Mbps (E1) for the Bosnia operations and fifty E1s for the Kosovo operations, totaling 225 Mbps of commercially leased bandwidth.

Operational Impact: ENDURING FREEDOM and IRAQI FREEDOM

In the aftermath of September 11, 2001, commercial SATCOM was used to augment MILSATCOM in USCENTCOM operations throughout Southwest Asia. Composite totals for the theater show that forty-two deployable Ku-band Earth terminals were used in support of warfighters amounting to more than 3.2 GHz of bandwidth on fifty-one different transponders. The significance of this bandwidth is reflected in the success commanders enjoyed in Afghanistan and the rapid advance of warfighters through Iraq. This commercial augmentation enabled forward units to communicate from day one with their headquarters, providing an overwhelming advantage to the U.S.-led coalition. One specific mission enabled by commercially provided bandwidth was the UAV reconnaissance of enemy positions. For the first time, real-time UAV video was available via multiple feeds: from the UAV to the ground station; from the ground station to the processing station; and from the processing station to forward planners. This synergistic capability allowed for real-time collaboration among warfighters, intelligence analysts, and the Pentagon, thereby establishing a new benchmark for the situational awareness, decision-making, and feedback loops. As Secretary of Defense Donald Rumsfeld noted,

> "linkages between UAVs, combat aircraft and bombers, and people on the ground, and the value that is created by those linkages...creates a very powerful effect."

Commercial SATCOM constituted more than 68 percent of the MILSATCOM used during Operation Iraqi Freedom. The dramatic growth in commercial SATCOM augmentation for deployed forces can be easily seen by comparing the relative amounts of military and commercial SATCOM used during Desert Shield and Desert Storm (75 percent military, 25 percent commercial) and during Operation Iraqi Freedom (32 percent military, 68 percent commercial). The growth in commercial SATCOM is not the result of a corresponding decrease in MILSATCOM. Rather, the use of both military and commercial SATCOM increased, further demonstrating the increased demand warfighters have for

satellite bandwidth and the way commercial SATCOM was able to meet those demands.

The low Earth-orbit satellite system, Iridium, made a different kind of contribution by providing communications with the use of handheld terminals from difficult environments. Iridium use grew from 17,255 calls of 54,755 minutes in September 2001 to 812,689 calls of 4,536,410 minutes in July 2003. During Operations Enduring Freedom and Iraqi Freedom, Iridium proved to be an excellent addition to the warfighters' communications tool kit. Its compact size and weight, secure capability, operational simplicity, and global coverage favorably positioned Iridium to supplement other tactical communications systems. These Iridium testimonials are also useful:

- "In this fast paced war, if a communications system was not functioning quickly, alternative methods were employed... The only systems consistently praised by the Marines were the Blue Force Tracker (SATCOM- though unsecure) and Iridium Phones (SATCOM). These systems provided reliable communications at all times. In many instances these systems were the sole means of communication." Marine Corps Systems Command Liaison Team, Central Iraq, May 2003
- "Only RELIABLE Communications out of the Valley (During Operation ANACONDA)" 75th Rangers
- "Without Enhanced Mobile Satellite Services (EMSS) we would have been unable to support the Air Mission in Afghanistan" USAF

- *"Iridium handheld satellite telephones with secure sleeves also proved to be invaluable for diverse SOF units conducting split operations in the rugged mountainous terrain. SOF liaison teams carried Iridium units during all operations with the Northern Alliance."* BG (P) James W. Parker, USA, Director, Center for Intelligence and Information Operations, SOCOM. (Signal Magazine, March 2003)

- *"We could go in there naked with flip-flops and as long as we have good radios, we could do our job."* Captain Jason Amerine, USA, 5th Special Forces Group. (Washington Post, 11 December 2001)

SUMMARY

In this paper, we have discussed the evolution of communications satellites from a paper concept in 1945 to the sophisticated systems currently in operation. The emphasis was on MILSATCOM, but many of the technological achievements were common to both commercial and military applications.

Most of the early commercial satellites served as relays between large, fixed Earth terminals that provided an alternative to submarine cables and landlines. The military required communications to be transportable or to be supported by mobile terminals. The UHF frequency band provided mobile services, but the data rates were very limited. The SHF band provided moderate data rates to vehicular and nontransportable terminals. Finally, as the technology became available at the EHF range, significant data rates became available to terminals mounted on High Mobility Multipurpose Wheeled Vehicles. Milstar changed the role of satellites from space-based relays to a space-based network. This evolution will continue in the future.

The implementation of these systems has required an enormous amount of technological development, from basic research to advanced development. Funding came from both government and commercial sources. The launch schedule for a particular satellite type may span a decade or more. Thus, delays may be significant as new technology is incorporated.

MILSATCOM played a vital role in all recent military operations. This paper provided some representative examples. This dependence on SATCOM will continue to grow.

APPENDIX: SATELLITE SYSTEM FUNDAMENTALS

Several basic ideas are fundamental to any discussion of satellites. A brief review of these ideas appears here.

Orbits

Most satellite communication systems use satellites in a geostationary orbit, as shown in Figure A.1. The basic idea is straightforward. The satellite is placed in an orbit that lies in the equatorial plane and is 35,786 kilometers above the surface of the Earth. The satellite is inserted into this orbit with a linear velocity such that its angular velocity is identical to the angular velocity of the Earth. At this altitude and velocity, the gravitational force and satellite's momentum are balanced so that the satellite remains in this orbit. In practice, other effects (e.g., eccentricity of the Earth, gravitational forces from the Sun and Moon) must be considered, so a process called station-keeping is required to maintain the satellite in this desired orbit. This orbit has the advantage that to an observer on Earth the satellite appears to be stationary.

In addition, the altitude is sufficiently high that messages can be exchanged by terminals separated by about one-third of the Earth's circumference. Thus, three satellites separated by 120° could provide global coverage between the latitudes of ±70°. Earth coverage contours for a geostationary satellite located above the equator in the middle of the Atlantic Ocean are shown in Figure A.2. The various curves correspond to the minimum elevation angles of the Earth terminals in the coverage area.

The disadvantage of the geostationary orbit is that the range causes significant signal attenuation (proportional to the square of the range) and a delay of about 300 milliseconds.

SATCOM Frequencies

A number of different frequency bands are allocated for SATCOM. Their characteristics are an important factor in the design of a particular satellite system. The allocations and usage for both commercial and military

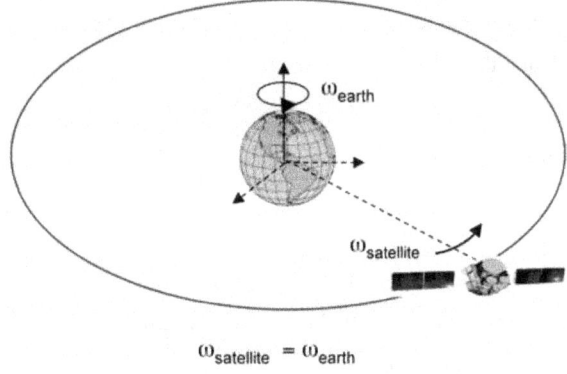

$$\omega_{satellite} = \omega_{earth}$$

Figure A.1. Geostationary Equatorial Orbit.

satellites are shown in Figure A.3.

In this paper we focus our attention on four frequency bands:

- UHF: 225–400 MHz
- SHF: 7.25–7.75 GHz, 7.9–8.4 GHz, 20.2–21.2 GHz
- Ku and Ka: 12–14 GHz, 30–31 GHz
- EHF: 43–45 GHz, 60 GHz

Two important factors are the location of the band in the frequency spectrum and the amount of bandwidth allocated. We discuss these factors for the above bands.

The UHF band has a wavelength of about 1 meter (exactly 1 meter at 300 MHz). This results in a propagation that is resistant to atmospheric effects, such as rain, and is able penetrate foliage cover. However, it is generally not practical to use antennas that focus the transmitted energy in a given direction. Thus, most UHF systems use antennas with modest gain and reasonably wide beams. This restricts the data rates that can be sent in UHF systems, but the small antennas make mobile and man-portable terminals feasible.

The bandwidth available at UHF is relatively small. This bandwidth limit has two effects. First, it restricts the data rates that can be sent through the system (typically, a digitized voice channel is 2,400 bps). The second effect is the inability to resist enemy jamming. Antijam techniques rely on either spreading the information over a large bandwidth to force the jammer to spread its power, or using a sophisticated antenna that can place a null in the direction of the jammer. Neither of these techniques are feasible at UHF.

The SHF band has wavelengths that vary

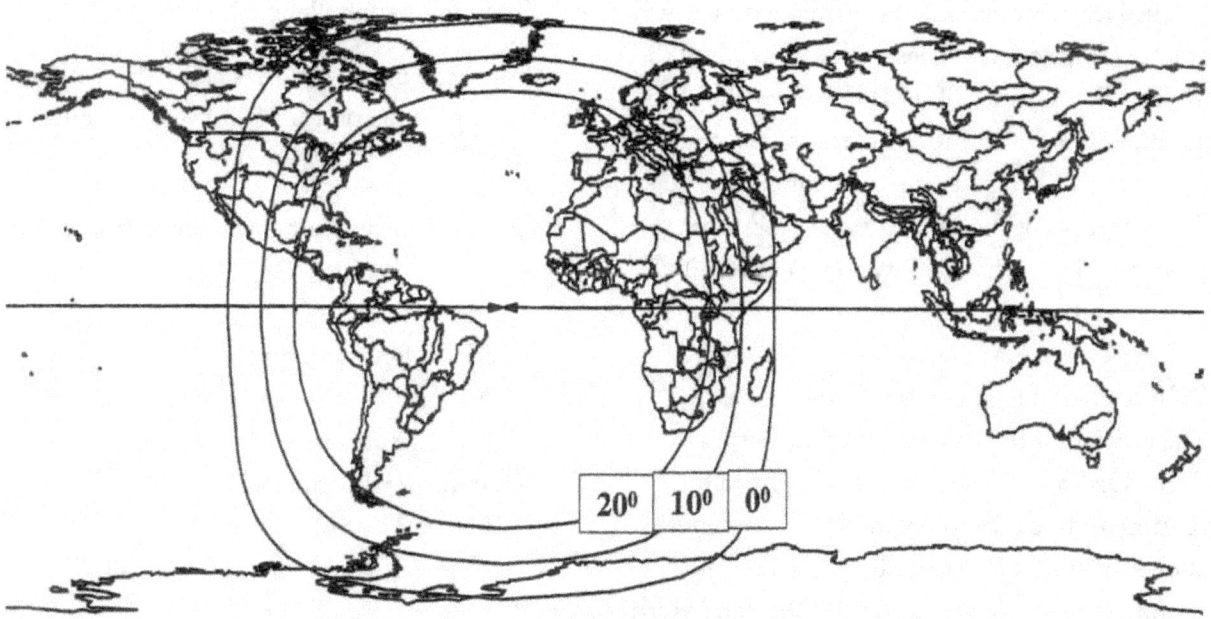

Figure A.2. Earth coverage contours for a geostationary satellite.

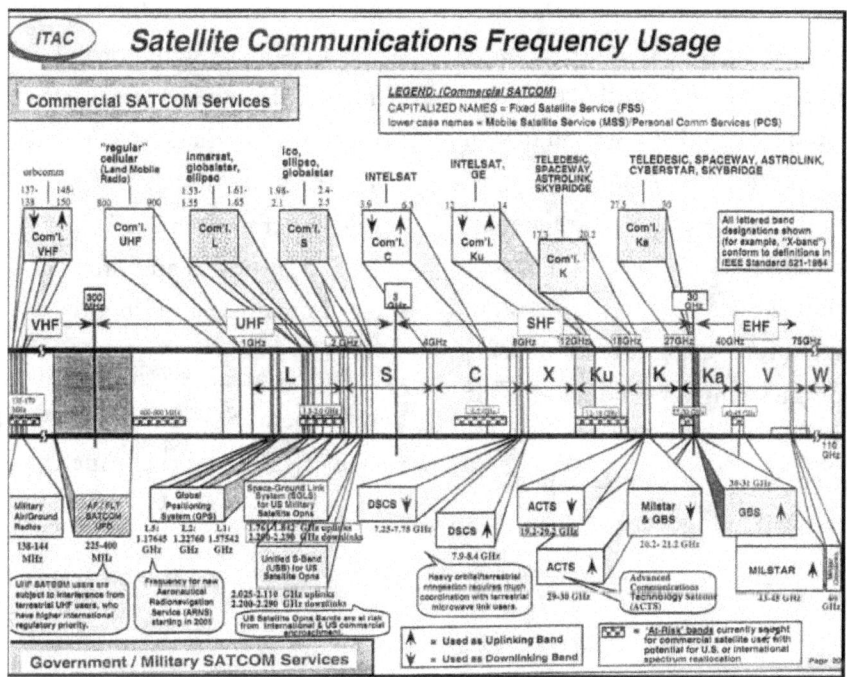

Figure A.3. SATCOM Frequency Usage.

from 10 centimeters at the lower end (3 GHz) to 1 centimeter at the upper end (30 GHz). SHF systems can utilize parabolic antennas where gain is proportional to the square of their diameter in wavelengths. Large antennas (40 feet to 60 feet in fixed systems, and 8 feet to 20 feet in transportable systems) can focus the energy in the desired directions and allow significant data rates.

The larger bandwidth allocated allows the use of antijamming techniques such as spreading the information across a 500 MHz bandwidth. In addition, it is feasible to utilize antennas on the satellites that can place nulls on jammers. The disadvantages of the SHF band are increased atmospheric effects, inability to penetrate foliage, and the size of the terminals.

The Ku and Ka bands are subsets of the SHF band and have similar characteristics. The portion of the EHF band of interest in SATCOM has wavelengths ranging from 0.67 centimeter to 0.5 centimeter. These wavelengths allow the usage of very small (1 to 2 foot) terminals that provide high gain. The bandwidth available allows for significant antijam protection.

The disadvantage of EHF is that atmospheric effects, such as rain, can cause significance attenuation of the signal, and the system design must take this possibility into account.

One characteristic of the different frequency bands that has become less important in the post Cold War era is the time it takes a system to restore adequate propagation after a nuclear explosion. These times range from days at UHF to minutes at EHF.

The characteristics of the various frequency bands are summarized in Figure A.4.

A.3. Technology Challenge

Some of the technologies that had to be developed to transition SATCOM from concept to reality are:

i. Launch vehicles to deliver the satellites into orbit
ii Satellite bus
  - Lightweight structure
  - Power generation (solar cells, batteries)
  - Orbit maintenance (thrusters)
  - Satellite stabilization
  - Thermal control
  - Telemetry, command, and control
iii. Communications payload
  - Antennas for receive and transmit
  - Low-noise amplifiers
  - On-board signal processing
ii. High-power transmitters (traveling-wave tube amplifiers)
iv. Earth Terminals
  - Antennas
  - High-power amplifiers
  - Low-noise amplifiers

All of these areas required significant research and development. Specific examples are highlighted in various sections of this paper.

Frequency Band	Bandwidth	Information Transmission Capacity	Atmospheric Effects	Scintillation Effects	Susceptibility To Jamming	Mobility
Ultra High Frequency (UHF)	Low	Low	Low	High	High	Mobile
Super High Frequency (SHF)	Moderate	Moderate High	Moderate	Moderate	Moderate	Transportable
Extremely High Freq. (EHF)	High	High	High	Low	Low	Man Transportable

Figure A.4. Characteristics of Frequency bands.

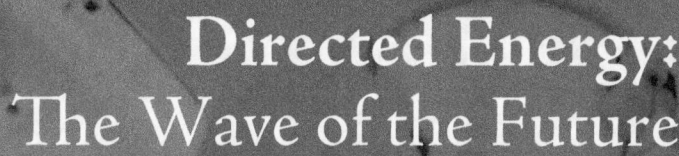

Directed Energy:
The Wave of the Future

Robert W. Duffner

Robert R. Butts

J. Douglas Beason

General Ronald R. Fogleman, USAF, Ret.

Abstract

In many ways the wave of the future for directed energy technologies began with the discovery of the laser in May 1960. A beam of energy that travels at the speed of light immediately appealed to the military community, which envisioned high-power lasers as a new class of weapons, destined to revolutionize the science and art of warfare in the twenty-first century. To turn that concept into reality, the Air Force laboratory system took the lead and maintained it in the development of an operational airborne laser system.

The focus of this paper is on the development and transformation of the airborne laser, covering nearly four decades as it evolved from a "laboratory in the sky" to a full-scale, operational weapons platform capable of unleashing deadly beams of light to disable or destroy ground- and air-launched missiles. Unmistakably, the Airborne Laser Laboratory ranks as the Wright Flyer of the laser world and has served as the technological bridge between laboratory research and the current Airborne Laser. Along the way, new and improved technologies emerged from the Air Force laboratories, such as the chemical oxygen iodine laser and sophisticated adaptive optics, which, combined, have enabled the propagation of a high-quality, high-powered beam through a turbulent atmosphere to ensure the long-range delivery of sufficient energy to engage and disable enemy weapons systems.

INTRODUCTION

At the height of the Roman Empire, Roman Legions armed with arrows, spears, swords, and shields devastated the more numerous but ill-equipped barbarian hordes. From the ancient world of Rome to the twentieth century, the means to wage war shifted dramatically from reliance on the physical prowess of the individual soldier to the modern weapons of warfare that depend on the scientific principles of physics, optics, chemistry, and other related disciplines. As the historical timeline of weapon development progressed over the ages, so did the level of precision and lethality of the emerging weapons. On 9 August 1945, a lone B–29 aircraft flying over Nagasaki, Japan, dropped a single atomic bomb that ultimately ended World War II. In February 1991, precision-guided smart bombs, ground-hugging cruise missiles, and invisible Stealth fighters forced the massively equipped Iraqi army to its knees.

Although decades apart, these overwhelming victories had one thing in common. They exploited technology to incur a revolution in military affairs that was so dramatic, so disruptive, and so profound that it changed not only the way wars were fought, but how nations interacted. Today, the next revolution in military affairs is about to begin. This revolution is not built on bombs or bullets, or anything you can hold in your hands. It is made of ordinary light in the same spectrum of energy found in your microwave, your light bulb, or in your television remote control; it's called directed energy. With recent advances in one form of directed energy called the chemical oxygen iodine laser (COIL), invented by Air Force researchers at what is now the Directed Energy

Directorate (DE) of the Air Force Research Laboratory (AFRL), and supporting beam-control technology such as adaptive optics, also advanced by Air Force researchers, the Airborne Laser (ABL) may constitute the key arsenal in this next revolution in military affairs.

The wave of the future for directed energy weapon technologies in many ways began with the discovery of the laser in May 1960. It was this unpredicted and unprecedented event that radically changed the thinking of how the U.S. military would apply this groundbreaking concept of directed energy to the development of operational weapon systems. The laser—he term derives from light amplification by stimulated emission of radiation—would move forward over the next forty-three years as one of the most promising technologies in the field of directed energy. Moreover, the United States made a substantial commitment to laser research because a beam of energy that could travel at the speed of light immediately appealed to the military community that envisioned high-power lasers as a new class of weapons destined to revolutionize the science and art of warfare in the twenty-first century.

HISTORICAL PERSPECTIVE

One of the first conceptual directed energy systems that emerged in the 1960s was the ABL. To turn that concept into reality, the Air Force laboratory system took the lead and maintained it as the most prominent Department of Defense (DoD) player that steadily pursued an extensive ABL research and development program. That work focused on the transformation of the ABL and its associated technologies for more than four decades. During that time, the ABL evolved from an experimental laboratory in the sky' to a full-scale, operational weapon platform capable of unleashing deadly beams of light to disable ground- and air-launched missiles.

The Airborne Laser Laboratory (ALL), a highly modified NKC–135 research aircraft (the equivalent of the commercial Boeing 707), was the first aerial platform to integrate a high-energy laser with a precision pointing and tracking system. Unmistakably, the ALL ranks as the Wright Flyer of the laser world; it is the technological bridge from early lasers and field demonstrations of lasers and beam-control components to the next-generation ABL, currently under development. Along the way, new and improved technologies emerged from the Air Force laboratories, such as the COIL and sophisticated adaptive optics that would become critical components of the ABL system. These technological advances proved invaluable in giving the ABL the capability to propagate a quality, high-power laser beam through a turbulent atmosphere to ensure the long-range delivery of sufficient energy to engage and disable enemy weapon systems.

In assessing the evolution of the ABL, one must keep in mind a number of important historical themes. To start with, the development and testing of the first ABL represented a significant turning point in the history of military science and technology. The ALL was truly revolutionary because of the results it achieved, which were completely beyond the capabilities of any other type of weapon system in existence. In 1983 the ALL projected a coherent beam of light from a moving aircraft to shoot down five supersonic AIM–9B "Sidewinder" air-to-air

missiles. This historical first was an event of major proportions that had far-reaching consequences on the future development of advanced laser weapons and directed energy.

A second theme is directly linked to the first. A distinguishing feature of the ALL was that it was a revolutionary weapon development program intended to radically change how people "thought about" fighting future wars. The ALL was not simply a first-of-a-kind event that occurred and then disappeared. Rather, the first ABL opened the door for DoD to make some risky decisions to move ahead to invest in a completely new class of directed energy weapons. The proof of that change in thinking is the second-generation ABL currently being built and scheduled to be deployed into the operational Air Force around 2008. Until recently, the ABL was the second-highest priority weapons acquisition program in the Air Force until the program transferred on 1 November 2001 to the Ballistic Missile Defense Organization, a joint service organization under DoD. (The F/A–22 tactical fighter is the number-one program.) To underscore the importance of the reality of lasers and their implications for the future, in July 1998 former Air Force Chief of Staff Ronald Fogleman confidently predicted, "Directed-energy weapons are going to be the centerpiece of the twenty-first century Air Force."[1]

The third theme has to do with people. No doubt, the ALL succeeded because of the joint effort among the military, civil servants, and contractors. What is unique about this particular program is that the military was truly a critical "hands-on player." Traditionally, the military manages weapons development programs. But in the case of the ALL, a highly educated and talented military group of scientists and engineers served as the grunts in the trenches and were intimately involved in working the day-to-day solutions to difficult scientific and technical problems.

Finally, it must be remembered, strong and steady leadership made a difference in the success of the ABL to elicit the best from a talented, but very diversified, workforce. Colonel Don Lamberson, working from the Air Force Weapons Laboratory (AFWL) at Kirtland AFB in Albuquerque, New Mexico, was the indisputable leader of the ALL program. Lamberson was a unique individual who possessed the rare combination of highly competent technical expertise coupled with extraordinary leadership and people skills. He was the one who conceptualized, sold, and led the ALL program. Motivating people and instilling in them a sense of urgency were two of his strongest traits. Lamberson was fond of saying in the early years of the ABL, "No one knew where the research would take us." In the end, it led to success.[2]

Dr. Petras Avizonis, Lamberson's chief scientist over the years, described his boss as simply different. "Lamberson was able to convey in his briefings his vision from a technical point of view, and his ability to articulate that technically in a way senior officers and DoD officials could understand it. He didn't talk down to them technically. So he had a fantastic combination of abilities, which was in large part the reason the Air Force for some ten years stood solidly behind the ALL demonstration." Lamberson had all the wrong credentials for making general. He never flew an airplane and he never commanded operational troops, yet he rose to become a major general primarily on the basis of his scientific talents and outstanding leadership record. As former Secretary of the Air Force Hans Mark put it in describing the success of the ABL, "Above all, leadership, persistence, and dedication prevailed and made the difference." The lesson learned was that people made a tremendous difference in revolutionizing science.[3]

ORIGINS

Although the Air Force would become the military leader in exploring the possibilities of using lasers on aircraft in the 1970s, the roots of the ALL program stretched back to research conducted by private industry a decade earlier. On 15 May 1960, Dr. Theodore Maiman, a senior scientist at Hughes Research Laboratory in Malibu, California, generated the world's first laser beam. Despite measuring only a few watts of power, the discovery of the ruby laser signified a revolution in the science of light that brought with it both confusion and promise for future applications. Maiman was one of the first to realize this and was cautiously optimistic about the significance of his groundbreaking research when he remarked, "The laser is a solution looking for a problem."[4]

However, the identification of problems and the applications of solutions proceeded at an extremely rapid pace. In the medical community lasers were used for delicate eye surgery to remove cataracts or to repair detached retinas. Commercially, lasers read bar codes on grocery items at the supermarket checkout counter and are just as capable of cutting through steel or drilling precision holes in the nipples of baby bottles.[5]

Maiman's invention quickly whetted the appetite of the defense community that envisioned high-power lasers not as a moneymaking enterprise, but as a new class of weapons that in the future would tip the strategic balance of power in favor of the United States in the ever-changing environment of the Cold War. The Pentagon wanted to build on Maiman's seminal work to develop a high-power laser, later defined as having a power output greater than 20 kilowatts, which would eventually be the unbeatable operational weapon. Ballistic missile defense and antisatellite and antiaircraft missions led the list of the military's applications for laser weapons.[6]

Military visionaries looked at lasers operating at the speed of light as a radical departure from the traditional kinetic energy type of weapon that relies on a projectile colliding with a target as the kill mechanism. In simplest terms, three conditions are necessary for lasing. First, some type of substance (gas, liquid, or solid) is needed to produce a beam, even Jell-O and bourbon whiskey can lase! Second, an intense energy source (electricity, a chemical reaction, a pulse-discharge lamp, etc.) is required to excite and alter the condition of the selected material. Third, a device commonly referred to as a resonator with mirrors at each end is required to extract the precise optical energy in the form of a beam.[7]

Although it was a complex process to generate a laser beam and to accurately direct it to a target, the military believed that in the long run the potential advantages outweighed the disadvantages of building a laser system. One major benefit is that light emitted by a laser is highly directional because its photons, or packets of light, are ""in step," precisely aligned in the same direction, accounting for the brightness of the beam. This allows the energy of the beam to be highly concentrated and focused on one spot so the beam penetrates the target quickly.[8]

Two other important factors convinced DoD officials that lasers were ideally suited as the ultimate weapon. Perhaps the most attractive feature of a laser beam is that it travels over long distances at the speed of light, about 186,230 miles per second. It takes only 6 millionths of a second (6 microseconds) for a laser beam to travel 1 mile. This means an operator does not need to lead the target because large amounts of energy can be delivered to a target nearly instantaneously. In addition, the lightning speed of the laser gives the operator more time between shots to identify and select other targets before committing to refiring the laser. Consequently, the target has virtually no time to evade the beam.[9]

Once it hits its target, the laser inflicts damage by rapidly heating and burning a hole in the target's surface, melting structural and electronic components, blinding sensors and detectors, and, in some cases, igniting on-board flammable materials. Lasers fired in short bursts (pulses) can also inflict damage from the explosive shock waves created by rapid heating of the target material by the beam.[10]

Besides the technical advantages, a second attractive feature of lasers was driven by political considerations. Lasers are not weapons of mass destruction like nuclear bombs and missiles. Rather, they are highly selective because all their destructive power is concentrated on one small area, which reduces the potential for civilian casualties as well as produces less collateral damage to structures around the target. The notion that a laser could be used as a "surgical scalpel" was very attractive to many of the top decision makers at DoD and was one of the main reasons for increased congressional funding for laser research in the 1970s.[11]

Recognizing the potential military payoffs for lasers, DoD's Advanced Research Projects Agency (ARPA) was the first government organization to sponsor laser research and development work. The decision to create ARPA on 7 February 1958 was a direct response to the Soviet launch of Sputnik on 4 October 1957. ARPA recruited the best scientific and engineering minds in the country to conduct high-risk research that would lead to rapid technological breakthroughs of a "revolutionary nature," in contrast to the slower, more traditional "evolutionary" approach to weapon development.[12]

At Kirtland AFB, the Air Force Special Weapons Center (AFSWC) was able to secure $800,000 from ARPA on 26 February 1962, which was the Air Force's first shaky step into the world of laser exploration. This money and follow-on funding were used primarily to conduct vulnerability studies to measure how much damage a laser could inflict on various materials. By the late 1960s, the AFWL, which had been carved out of AFSWC in 1963, had earned the reputation as the center of excellence for laser research in the Air Force.[13]

In July 1963 the AFSWC presented its findings from its investigation of glass ruby lasers. Results from this work were very disappointing because enormous devices would be needed to reach high power, and heat buildup with glass lasers severely distorted the beam.

As a young captain, Don Lamberson was working on this project and explained, "We showed conclusively that you really just couldn't get there." However, the Air Force refused to give up on perfecting a working laser.[14]

Two pivotal technical advances changed the outlook for the future of lasers in a very positive way. A Bell Telephone scientist, Dr. Charles Kumar N. Patel, discovered in April 1964 that molecular gas carbon dioxide (CO_2), through electrical pumping, could be used as a laser medium. This was the first continuous-wave CO_2 laser. His breakthrough touched off widespread interest in CO_2 lasers, mainly because of their promise to attain higher power levels and efficiency.[15]

A second major milestone occurred in 1966 when Dr. Edward Gerry of AVCOEverett Research Laboratory demonstrated a new pumping technique to stimulate population inversion (the phenomenon that produces a laser beam) by rapid expansion of a hot equilibrium gas mixture through a bank of supersonic nozzles. This was the first gas dynamic laser. AVCO continued to refine its gas dynamic laser and produced 168 kilowatts in March 1968. This represented tremendous progress. Until 1964, power levels of lasers were so low that "Gillettes" defined as the number of razor blades a laser could penetrate in .05 second were used as the standard unit for measuring power. All these events—AFWL's initial laser research, invention of the gas dynamic laser, and AVCO's advances in achieving higher power levels—hastened the Air Force's interest in the military applications of gas dynamic lasers. That interest turned into a practical research and development program in 1968 when Headquarters, Air Force, authorized AFWL to proceed with building and testing a ground-based gas dynamic laser.[16]

AIR FORCE LEADS THE WAY

Once given the green light, AFWL's first priority was to design, build, and fire a CO_2 gas dynamic ground-based laser capable of engaging static and moving targets. AFWL vigorously pursued this goal using the tri-service laser (DoD gave each service an identical AVCO-built CO_2 laser for research purposes) at Sandia Optical Range (Sandia's name changed to Starfire Optical Range in the late 1980s) located on the southeast section of Kirtland AFB. Looking toward the future, AFWL recommended placing a laser on an aircraft to be tested as a possible tactical or strategic weapon. The Air Force had thus planted the seed that would grow into the development of the ALL, perfecting the physics and engineering requirements to enable firing a laser from an airborne platform. But it was clear to everyone that the laser would first have to be demonstrated on the ground before it could be integrated and tested on an airborne platform.[17]

The ground-based laser experiments at the Sandia Optical Range were part of Eighth Card, the code name of a highly classified DoD program to advance laser technology quickly. The name Eighth Card originated from seven-card stud poker, implying the advantage went to the player who held the extra eighth card. Translated to the political arena, this suggested that if the United States held the eighth laser card, it would hold a distinct advantage over the Soviets in the Cold War or in any future military confrontation.[18]

The first step to prove the feasibility of a ground-based laser occurred in October 1971 when the field-test telescope (built by Hughes Aircraft) was combined with a CO_2 laser and referred to as Air Force Laser I. Mating of these two critical system components resulted in the telescope's focusing the low-power (1,000 watts) laser on a target pod mounted on a T39 aircraft flying downrange. Although the T39 suffered no damage because of the low power of the beam, this was a tremendously important event because it was the first time a laser beam had hit an aircraft in flight. In December 1972, the Air Force reached a second major technical milestone when it successfully directed and focused a high-power laser beam (150 kilowatts) on a rotoplane located 1,760 meters downrange from the laser device. Resembling a windmill contraption, the 30-foot-long arm of the rotoplane rotated a small target, the size of a postcard, 360 degrees along its vertical plane at 25 revolutions per minute.[19]

Project DELTA (Drone Experiment Laser Test and Assessment) served as the final climax to the ground-based program. On 14 November 1973, AFWL used a CO_2 laser to shoot down a radio-controlled aerial target (drone) as it flew a racetrack pattern between Sandia Optical Range and the Manzano Mountains. This was the first time in the history of the world that a high-energy laser had disabled an aerial target, an enormously important technical accomplishment, considering that lasers had been on the scene a mere thirteen years. Equally important, this historic event served as a political rallying point to sustain laser work in general. More specifically, DELTA helped persuade the purse-string holders and the doubters, unfamiliar with the new technology of lasers, of the urgency to move forward with the ABL program.[20]

In January 1973, General Dynamics had extensively modified the NKC–135 research aircraft to serve as the first ABL testbed. This ALL was intended not to be a prototype but rather a science laboratory in the sky for proving the physics of lasers by conducting a three-stage research and test program. Cycle I was to demonstrate that the airborne pointing and tracking system could accurately track a maneuvering target when the ALL was airborne. This was accomplished through a series of flight tests that ended in November 1973, with no laser in the airplane, over White Sands Missile Range in southern New Mexico.[21]

Cycle II mated a low-power electric discharge laser with the airborne pointer and tracker to determine if the beam could be directed from the turret of the ALL aircraft to intercept an aerial target. By 1976, after more than 100 flight tests, the ALL proved it could track a diagnostic NC–135 aircraft that flew alongside the ALL and, at the same time, aim a beam from the ALL with sufficient precision to send it through a small window on the side of the diagnostic aircraft. This was a milestone of major proportions because it was the first time in history a laser beam had been fired from an airborne platform to intercept a flying target.[22]

Although armed with a high degree of confidence derived from the success of Cycles I and II, the ALL team realized Cycle III would be an ambitious undertaking. A diverse group of workers now had to face an enormously complex engineering project to design, build, and integrate a vastly more

powerful laser with a water-cooled, optical beam-control system. Integrating and perfecting the laser system during Cycle III was both frustrating and rewarding for those who worked on the ALL. However, after more than seven years of extensive ground-testing and flight-testing—plain hard work marked by numerous failures for every success—the ALL was ready to face its greatest challenge.[23]

On 26 May 1983, the ALL made history again when it completed its first successful engagement of an aerial target and destroyed an AIM9B Sidewinder air-to-air missile over the Naval Weapons Center Test Range at China Lake, California. The beam remained on the nose of the missile long enough (4.8 seconds) to heat up and damage the sensitive components of the guidance system, causing the missile to break lock. Colonel John Otten, the ALL test director, described that history-making first shootdown as "if someone wrote the textbook. Tracking was rock solid, the beam remained on the nose of the missile to cause significant damage causing the missile to veer off course and crash." On four other airborne tests that followed, the ALL's precision pointer and tracker succeeded in disabling all four Sidewinder missiles. Hailed in the scientific community as an exceptional step forward in the world of directed energy, these demonstrations proved for the first time that a high-power ABL could intercept and destroy an air-to-air missile.[24]

Four months later in the fall of 1983, the beam intercepted three Navy BQM–34A drones over the Pacific near Point Mugu, California. For the first drone, the beam slightly missed the fuel tank aim point and did not cause the tank to rupture. On the second drone experiment, the beam deposited sufficient energy to cause structural damage to the wing root, but the drone continued flying. For the third and final drone, the beam burned through the flight control box and melted numerous wires, which caused multiple circuit failures. As the electrical system failed, the drone went out of control, making a hard 90-degree roll to the right. It abruptly made a sharp pitch-down maneuver and crashed into the ocean. That was a confirmed kill.[25]

The ALL was a prime example of advanced science and technology that served to strengthen this nation's overall military strategy during the Cold War. Although it never became an operational weapon, the ALL certainly made substantial progress in developing new technology that offered to revolutionize the art and science of war. U.S. and Soviet nuclear deterrence based on the delivery of nuclear warheads launched from long-range bombers, submarines, and land-based missile silos provided the foundation of deterrence. The Soviets and Americans both realized the development of laser weapons was a very real threat that could easily disrupt the delicate balance of power in an unstable world. They also knew whoever could harness and apply the new technology of lasers first would be in a much stronger bargaining position to tip the scales of strategic power in their favor. Looking toward the future, the scientific and technical bricks and mortar of the ALL doubtlessly laid the foundation for the second-generation ABL currently under development.

ADAPTIVE OPTICS

As work progressed on the ALL, scientists at the AFWL pursued laser research programs independent of the ALL program. Results from those scientific investigations led to groundbreaking technological advances in two extremely important areas. One was the growth of adaptive optics; the other, the discovery and development of COIL. These technologies would become the two critical components of the new ABL system designed to shoot down enemy missiles in their boost phase.

The Air Force has a rich heritage spanning three decades in advancing adaptive optics technology, one of the essential enabling technologies for the ABL. One of the reasons attention turned to adaptive optics research in the 1970s was because the ALL mirrors were unsuitable for use on the next-generation ABL. The ALL water-cooled mirrors were not optimally efficient, were not reliable for transmitting a quality beam through the atmosphere, and were subject to corrosion that could damage their surface. Indeed, adaptive optics research at the AFRL and at its follow-on organization, Phillips Laboratory, led to many of the most important breakthroughs in the field.

Laser beams are electromagnetic waves, and an electromagnetic wave can be described in terms of its amplitude and phase, or wavefront. Just as a distorted wavefront can degrade the performance of an imaging system (think of a picture taken with a camera that is out of focus), a distorted wavefront can limit the intensity of a high-energy laser (HEL) at the target and prevent the desired target destruction.

By altering the wavefront of the HEL, an adaptive optics system can correct or compensate for the deleterious effects of beam aberrations to enable an intense and lethal concentration of laser energy at the target. The ABL uses two types of active optics to correct the HEL wavefront: fast-steering mirrors (FSMs) and deformable mirrors (DMs). An FSM is just a flat mirror that can be tilted rapidly about two axes to adjust the direction in which the HEL is pointed. Coarse pointing of the HEL is accomplished by directing a 1.5 meter telescope, but compensation for high-bandwidth tilt errors, such as those induced by acoustic disturbances within the aircraft, requires more agility than can be achieved by pointing the telescope. A DM is just what the name suggests: a mirror whose reflective surface can be deformed. The DM used by the ABL has hundreds of actuators attached to the backside of the mirror which can be electronically commanded to adjust the shape of the mirror's front surface and thereby adjust the wavefront of the HEL that reflects from it. Figure 1 illustrates the use of a DM to correct an aberrated wavefront.

The requirement for adaptive optics for the ABL arises from two sources: distortions imparted to the HEL onboard the ABL aircraft, and aberrations acquired by the beam along the propagation path through the atmosphere from the ABL to the target. The ABL COIL gain medium consists of gases flowing at supersonic velocities. Inhomogeneities within the medium can warp the wavefront of the beam that the gain medium produces. Also, even though the ABL optics absorb only a tiny fraction of the laser energy, the HEL is so powerful that the optics will still heat up and deform. The resulting changes in the surface figure of the optics are small, but uncorrected optics deformations of even a fraction of a micrometer can have a devastating impact on

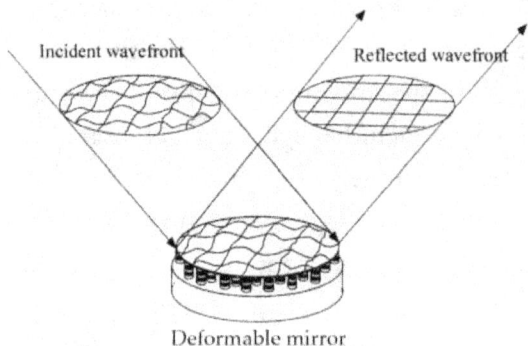

Figure 1. A deformable mirror is used to correct an aberrated wavefront.

system performance. Therefore, there must be a "local loop" compensation system onboard the ABL to correct these aberrations.

After the beam exits the nose of the ABL, it initially encounters airflow around the aircraft as well as turbulence that exists in the free atmosphere between the aircraft and the target. Light propagated through the atmosphere is always affected by turbulence, the term used to denote random localized fluctuations in the temperature and density of the air. The twinkling of starlight is perhaps the most widely known manifestation of the effects of atmospheric turbulence on light propagation. The laws of physics dictate that turbulence can severely impact the performance of imaging systems and HEL weapon systems. For example, astronomical telescopes must be large enough to collect sufficient light to detect dim objects. Without adaptive optics, atmospheric turbulence limits the resolution of an image produced by a large telescope—the ability to distinguish the stars in a binary pair—to a level of quality no better than that of a small

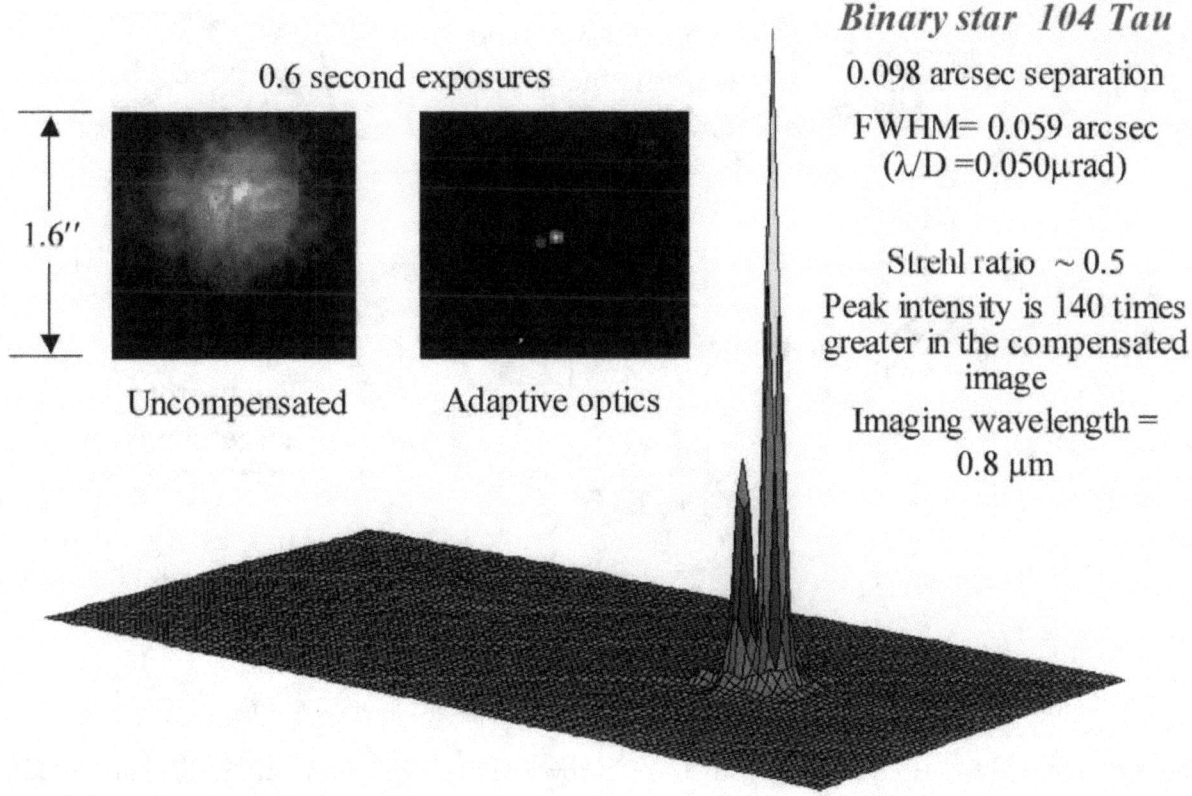

Figure 2. Binary star images taken with the AFRL Starfire Optical Range 3.5m telescope and adaptive optics system at Kirtland Air Force Base in New Mexico.

amateur backyard telescope. Figure 2 illustrates the impact of turbulence on astronomical imagery and the capability of adaptive optics to dramatically improve image quality.

A highly simplified depiction of the ABL adaptive optics system is shown in Figure 3. This system has two separate adaptive optics subsystems: a local loop system and a target loop system. The local loop system is used to correct the aberrations imparted to the beam in its generation and propagation inside the ABL aircraft up to the aperture-sharing element, which in the case of the ABL is a simple beam splitter. The target loop system corrects the aberrations imparted to the beam on its journey from the aperture-sharing element to the target.

In 1953, Horace Babcock first proposed using adaptive optics[26] for real-time compensation of astronomical imaging to create higher resolution by using large telescopes to their full potential. However, the approach he proposed was never implemented. A few years later R.B. Leighton[27] used the simplest form of real-time adaptive optics consisting of simple tip-tilt correction with an FSM to stabilize the images of planets during camera exposures and to improve image sharpness. Deformable mirrors, the critical element of any adaptive optics system, would not be developed for many more years. National defense requirements drove much of the progress on deformable mirrors in the late 1960s and into the 1970s. DoD

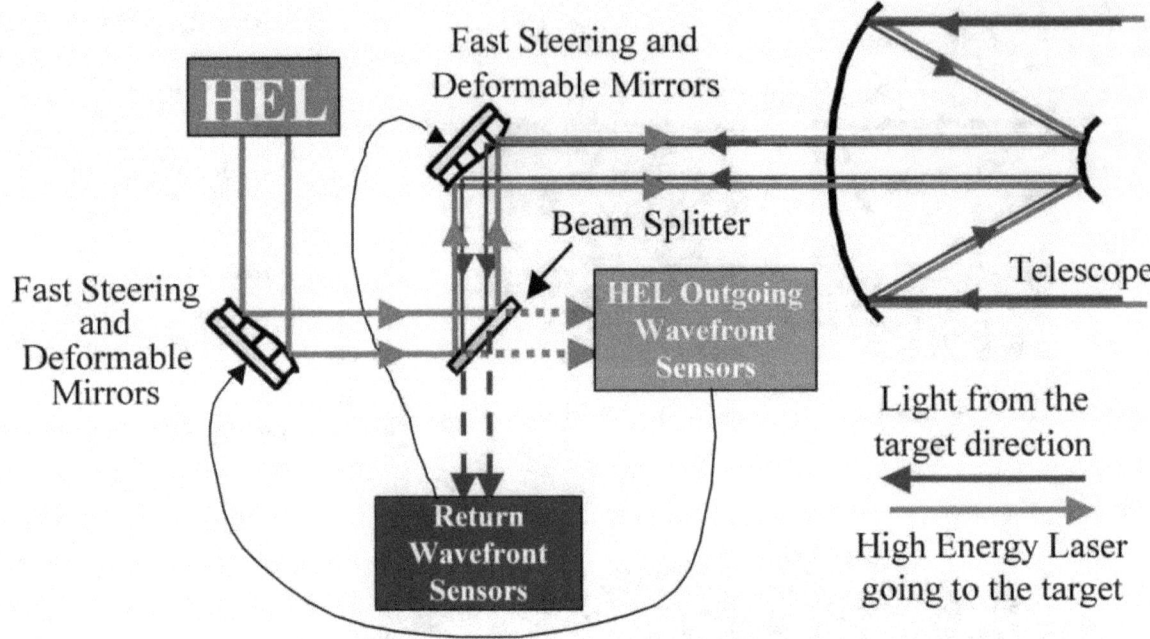

Figure 3. The ABL adaptive optics system compensates for both on board and atmospheric disturbances.

became interested in adaptive optics for their potential to improve the performance of imaging systems and of possible future laser weapons.

Rome Air Development Center and its contractors from Itek Corporation accomplished the first real-time, higher-order compensation of atmospheric turbulence using a deformable mirror in 1974.[28] (Rome Air Development Center later became Rome Laboratory and is now a part of the Information Technology Directorate of AFRL.) Rome personnel developed the real-time atmospheric compensation system and demonstrated the correction of a 632.8-nanometer laser beam propagating over a 300-meter horizontal path to a 30-centimeter receiving aperture. Their system used a monolithic piezoelectric mirror with twenty-one actuators, a white-light shearing interferometer wavefront sensor (WFS) with sixteen subapertures (areas within the aperture

over which the local wavefront tilt is measured), and a high-speed controller based on an analog computer. Both the monolithic piezoelectric mirror and the shearing interferometer represented breakthrough advances in the state of the art in adaptive optics hardware.

The first real-time compensation (beyond tilt) of astronomical images was reported in the literature in 1977.[29] The apparatus had a 30-centimeter receiver and adjustment was accomplished with six movable mirrors. Also in 1977, compensated images of artificial satellites were collected at Maui with an adaptive optics system mounted on a 1.6-meter telescope by the Advanced Radiation Technology Office of the AFWL which later became part of the Air Force Phillips Laboratory and then part of the Directed Energy Directorate. The system used a 37-actuator mirror driven modally to correct for focus and astigmatism while an

FSM compensated wavefront tilt. Both tilt and higher-order aberrations were measured by a six-subaperture WFS. For classification reasons, the results were not reported in the open literature, but details of the digital control architecture can be found in Corsetti et al.[30] Unfortunately, the poor sensitivity of the system limited its application to only the brightest space objects.

Rome Laboratory installed the first practical adaptive optics system, the compensated imaging system, on the 1.6-meter telescope on Mt. Haleakala at the Defense Advanced Research Projects Agency (DARPA) (ARPA's name changed to DARPA in 1972) Maui Optical Site in 1982.[31] The system pictured in Figure 4 used a DM with more than 100 actuators. A workhorse system, it provided compensated images of space objects continuously until the early 1990s.

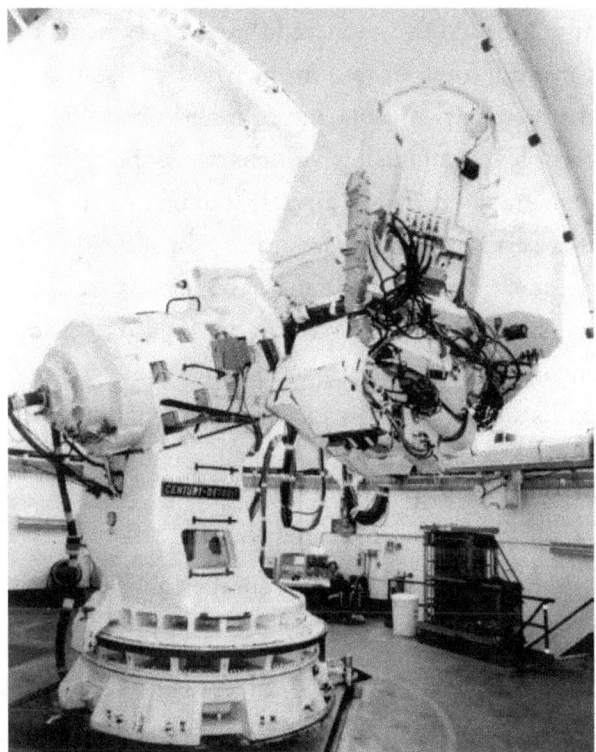

Figure 4. Compensated Imaging System mounted on 1.6 meter telescope on Mt. Haleakala.

During the 1980s, work continued to improve the technology for DMs. John Kenemuth and his group at AFWL sponsored DM development at Hughes, Itek, Perkin Elmer, and United Technologies. Much of the effort was devoted to improvements in linearity of response and reliability and to work on cooled DMs for HEL applications. The cooled DM technology has since been somewhat overtaken by events as dramatic improvements in ultralow-absorption coatings have largely eliminated the need for cooled optics. All ABL optics are uncooled. Advantages of uncooled optics include reduced weight, elimination of potential damage from coolant leaks, and, most importantly, the elimination of the HEL jitter caused by coolant flow and the coupling of pumps through the plumbing into the optics. AFRL developed many of the advances in optical coatings.

Development of artificial beacon technology was another important AFWL contribution to adaptive optics. Early adaptive optics systems, such as the compensated imaging system, relied on light from the object to provide the reference for the WFS. Usually, most of the available light was required for the adaptive optics system, leaving little signal for forming the image, and dim objects provided insufficient signal for effective compensation. Artificial guidestars created by backscatter of pulsed beacon lasers from the atmosphere were proposed as a solution to this problem. An AFWL team led by Dr. Robert Fugate and the Massachusetts Institute of Technology's Lincoln Laboratory performed the seminal experiments that proved the viability of this technique.[32] DoD's artificial beacon research was classified until May 1991 when DoD decided to declassify its artificial guidestar research, which was revealed at the American Astronomical Society's annual meeting in Seattle. Transfer of this technology to the astronomical community has had a revolutionary impact on the subsequent development of large ground-based observatories. Today, every large astronomical telescope being built incorporates adaptive optics in its design.

When the ABL was proposed in the early 1990s, it was understood that the problem of compensation of turbulence effects would be substantially different than it would be for ground-based imaging and laser transmission systems for which light propagation through the turbulence is nearly vertical. For ground-based systems, turbulence is concentrated within a few kilometers of the receiving or transmitting aperture. The ABL HEL, however, must propagate for hundreds of kilometers horizontally to destroy its target. Furthermore, because ground-based applications were dominated by turbulence at low altitudes, there had been less attention paid to the nature of turbulence at ABL operational altitudes at 40,000 feet. One of the key questions at the inception of the ABL program was whether or not the turbulence at high altitude would allow the concentration of lethal levels of laser energy on the target, even with adaptive optics. The Air Force laboratories performed research to answer that question and to better understand the fundamental phenomenology of high-altitude turbulence.

The strength of the optical effects of turbulence can be related to the magnitude of the temperature fluctuations in the atmosphere. The Air Force Geophysics Laboratory, now part of the AFRL Space Vehicles Directorate,

pioneered balloon-borne high-bandwidth-sensitive temperature probes that could be used to measure turbulence at altitudes up to 30 kilometers. Techniques they developed were used to characterize turbulence at several sites including Maui and White Sands Missile Range. The Geophysics Laboratory developed models of turbulence from these data, including the Clear1 model[33] that was adopted as the basis for the turbulence requirements for the ABL program. The balloon probes provided vertical slices through the turbulence profile. To obtain high-resolution data on the horizontal structure of turbulence, AFWL's DE developed techniques to measure the strength of turbulence using temperature probes mounted on aircraft.[34]

Efforts to characterize high-altitude turbulence for the ABL program culminated in the ABLSTAR experimental campaigns in 1999 and 2000.[35] DE conducted three deployments in three different seasons in the Middle East and the Korean peninsula and collected atmospheric turbulence data using balloons and a specially instrumented C–135 aircraft (dubbed Argus) that served as a flying laboratory. In addition to the in situ temperature measurements, a stellar scintillometer was also flown on Argus to sense the integrated effects of turbulence along slant paths similar to those of some ABL engagement scenarios. These experiments provided further validation of the models of turbulence used to specify ABL system requirements.

AFRL also conducted a number of propagation experiments to determine the efficacy of adaptive optics for the ABL application. In 1993, AFRL/DE conducted the Airborne Laser Experiment (ABLEX) that propagated a laser from one aircraft to another at separations up to 200 kilometers and at ABL operational altitudes. By measuring the irradiance distribution of the beam across an 80-centimeter aperture on the receiving aircraft, the performance with perfect phase compensation (perfect adaptive optics) could be calculated without making any assumptions regarding the structure of turbulence.[36] Real adaptive optics systems do not achieve perfect compensation, of course, but the experiment established that the physics limit of phase compensation with high-altitude turbulence was as expected. The physics limit for phase-only compensation was determined to yield Strehl ratios from 0.7 to 0.8. Those were extremely good measurements, considering a Strehl ratio of 1.0 represents perfection.

ABLEX was followed by the ABL Extended Atmospheric Characterization Experiment (ABLE ACE) in 1995.[37] ABLE ACE was another set of experiments that used measurements of a laser beam propagated between two aircraft flying at high altitudes, but it included a much richer suite of instruments than ABLEX. High-resolution temporal and spatial measurements of the laser beam's amplitude and phase were recorded on flights in the continental United States and in Korea. ABLE ACE was designed to be sensitive to every turbulence effect that was thought to be important for ABL performance. It replicated, at higher fidelity, the measurements taken in ABLEX and added high-bandwidth scintillometry, measurements of the phase difference between two beams propagated between the transmitting and receiving aircraft, as well as WFS data, full aperture wavefront tilt, and far-field imagery. After an extensive effort to understand all the experiment noise sources, all the data presented a consistent picture that provided even stronger validation of the models used to predict ABL system performance.

Parallel to the efforts to better understand high-altitude turbulence, DE conducted propagation experiments to determine the performance of adaptive optics compensation for turbulence distributed along the entire propagation path. Those experiments used a combination of range, laser wavelength, aperture size, and turbulence strength so that they were scaled to ABL scenarios, that is, the turbulence effects were identical to those that would be expected for the ABL. The Horizontal Propagation Experiment (HOPE), performed at the Starfire Optical Range in 1992, successfully used adaptive optics to compensate a visible wavelength laser propagating over a two-mile path. The experiment showed that adaptive optics significantly improved performance, but as understanding of compensation of distributed turbulence improved, much better performance was achieved later in similar experiments at the Lincoln Laboratory Firepond facility and at North Oscura Peak, located on the White Sands Missile Range.

Some of the early propagation experiments used a point-source beacon as the reference for tracking and adaptive optics. Of course, the ABL cannot rely on that degree of cooperation from potential adversaries who will not put beacons on their targets. The ABL must provide its own beacons, which it does by illuminating the missile target with separate illuminator lasers for tracking and adaptive optics (called the TILL and BILL, respectively). The ABL tracks the reflected TILL light and

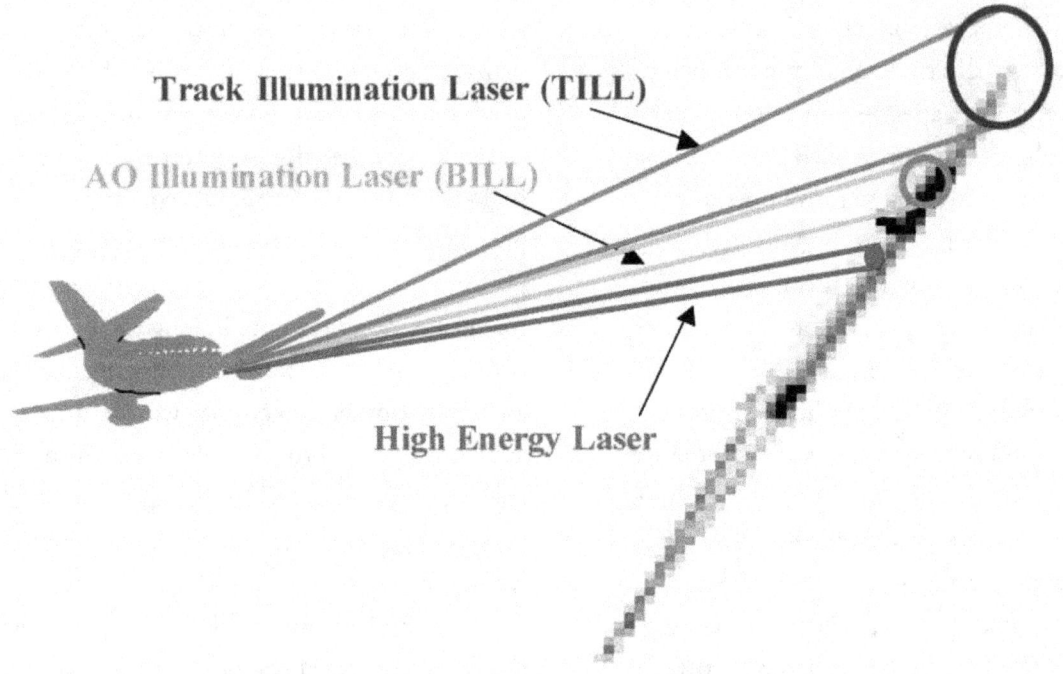

Track Illumination Laser (TILL)

AO Illumination Laser (BILL)

High Energy Laser

Figure 5. The Airborne Laser must illuminate the target with illuminator lasers, separate from the COIL high-energy laser, to provide signal for tracking and adaptive optics sensors.

sends the reflected BILL light to the WFS. This approach is illustrated in Figure 5.

In 1999, DE performed an important series of tests identified as the Dynamic Compensation Experiment (DyCE) at North Oscura Peak. A compensated surrogate HEL, a low-power laser used to score the experiments, was propagated to a diagnostic target board on an aircraft in flight at ranges of 30 to 50 kilometers. A laser beacon on the aircraft provided the reference for the tracking and adaptive optics. Although the target was cooperative, the tests used a dynamic target. In 2000, the Non-cooperative Dynamic Compensation Experiment (NoDyCE) followed DyCE. In NoDyCE, the important features of the ABL compensation architecture were replicated at the North Oscura Peak site.

The tracking and adaptive optics performed used reflected light from an illuminated target board on a Cessna aircraft. Performance of the system met all expectations. Figure 6 shows a comparison between the performance with the high-bandwidth tracking and closed adaptive optics loops, and performance with only low-bandwidth tracking and no adaptive optics.

CHEMICAL OXYGEN IODINE LASER

Shortly after the invention of the laser, DARPA recognized the potential utility of using lasers on the battlefield. As speed-of-light weapons, lasers are able to deliver focused, coherent energy at near-instantaneous velocities of over 186,000 miles a second (around Mach 88,000). Furthermore,

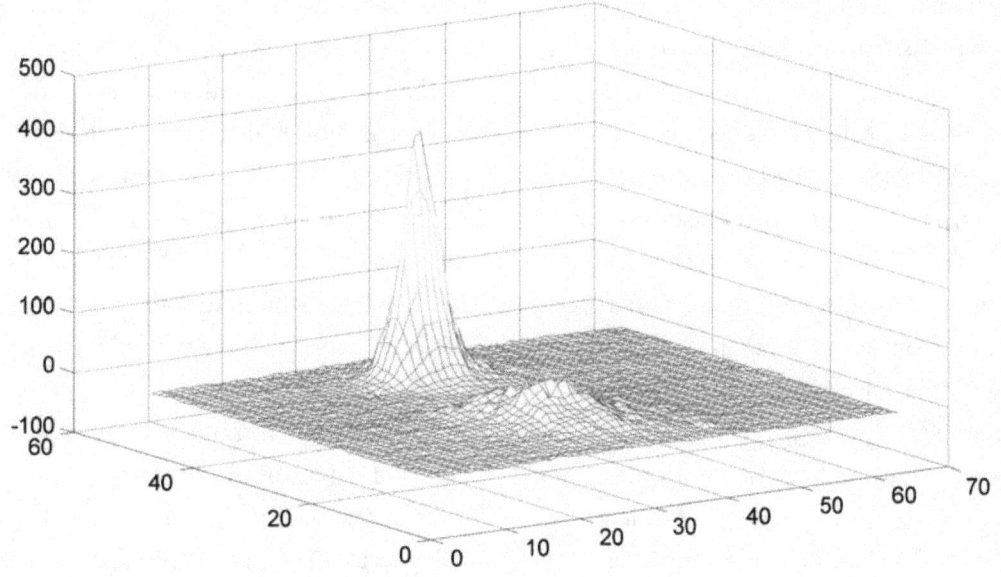

Figure 6. Average scoring laser profiles at the aircraft mounted target board recorded during one of the NoDyCE tests. The profile in the front was recorded without the adaptive optics loop closed, but with a low bandwidth (12 Hz) track loop using a cooperative beacon. The more intense profile was recorded with the adaptive optics and high bandwidth track loops closed. Loops were closed using reflected illuminator laser light from the non-cooperative target.

lasers are not constrained by ballistics, allowing warfighters the unprecedented opportunity to deliver effect on demand.

DARPA's early investments in advancing laser technology resulted in the hydrogen fluoride (HF) and deuterium fluoride (DF) lasers, the first high-power systems apart from the "traditional" CO_2 chemical laser. Although these various lasers worked well, they lased in relatively long wavelengths, requiring large optics and enormous infrastructure to store the energy necessary for their operation. And even though the CO_2 laser had been used in the ALL, the dream of one day putting an operational high-power laser onboard a plane, perhaps ultimately in space, would never be achieved unless a more efficient, shorter wavelength laser

could be found. Also, chemical lasers promised significant weight savings, an important consideration if the device were to be sent aloft.[38]

In the mid-1960s, Jerome V.V. Kasper and George Pimentel demonstrated the first successful chemical laser and the first iodine laser in experiments performed at the University of California, Berkeley. Unlike lasers that require electricity or flash lamps to generate a reaction, chemical lasers rely on chemical reactions to excite the molecules for light amplification (the *la* of laser) to create a beam of light.[39] Operated at a wavelength of 1.315 microns (a micron is equivalent to a millionth of a meter, a distance a thousand times smaller than a millimeter), the iodine laser appeared to be an ideal wavelength for airborne applications.

Specifically, it was not readily absorbed into the atmosphere, and its short wavelength would allow the use of smaller optics and lighter supporting infrastructure. If a high-power, continuous beam iodine laser could be invented, it would solve most of the problems associated with other high-power laser systems.

However, the Berkeley iodine laser operated only in short, single pulses and had nowhere near the power to use as a weapon. If this tantalizing iodine wavelength was to be used in the real world, it had to be generated by an entirely different method. In addition, it had to have the ability to "scale," that is, have the potential to increase in power by over a billion times, and it had to be a continuous beam instead of a short pulse. An analogy is that although the Wright Flyer and the F/A–22 Raptor both fly in air, they are worlds apart in performance capabilities. The same advancements were needed if the iodine laser was ever going to be a serious contender as a directed energy weapon.

Around the same time that the short-pulse iodine laser was invented, the Air Force introduced extremely high-flying reconnaissance aircraft such as the U–2 and the SR–71 into the operational force structure. These and other high-flyers flew in the stratosphere, where knowledge regarding the chemical properties of the air was wanting. Consequently, there were concerns about the possibility of the effects of air pollution and corrosion on aircraft at those altitudes. As a result, the Air Force Office of Scientific Research (AFOSR) in 1965 awarded a contract to Dr. Elmer A. Ogryzlo at the University of British Columbia to study the reactions of a particular chemical state of the oxygen molecule, the singlet delta oxygen, which was present in the upper atmosphere. With his graduate student S.J. Arnold, Ogryzlo noted a strange chemiluminescence when mixing singlet delta oxygen and iodine.[40]

During a sabbatical at Cambridge in the late 1960s, Ogryzlo worked with collaborator Brian Thrush. It would be Thrush's graduate student R.G. Derwent who first noted the emission of 1.315-micron light from the iodine atom while studying the spectroscopy of an oxygen iodine system. Derwent and his colleagues published a series of three landmark papers detailing the dissociation, or breakup, of elemental iodine (I_2) by the excited state of oxygen known as singlet delta oxygen. They thus discovered a chemical excitation mechanism of the iodine atom. Although Derwent and Thrush first suggested the inversion or alteration of iodine if enough singlet delta oxygen could be produced (which became the central problem), it is generally agreed Ogryzlo first observed the phenomena that established the theoretical groundwork for Kasper and Pimentel's invention of the chemical excitation of the iodine laser.[41]

The publication of Derwent's work in 1971 provided the initial impetus for researchers at the AFWL, where senior scientists Dr. Petras Avizonas and Lieutenant Colonel Mel Bina took an avid interest in Derwent's work. In 1973, Dr. Alan McKnight began working on the oxygen iodine laser at AFWL with the lab's first direct funding. McKnight's initial work consisted of conducting a literature search for developing the kinetic rate constants necessary to computationally model an oxygen iodine laser. Before an attempt was made to build this laser, a massive effort was undertaken to verify the kinetics involved in the production of inverted or altered iodine. Plus, McKnight sought to locate a source to generate the critical singlet delta oxygen, first unsuccessfully trying microwave-excited oxygen.[42]

In-house AFWL researchers, consisting of military and civilian scientists, worked with collaborators to optimize the production of singlet delta oxygen. In 1973, McKnight approached Major Bill McDermott, a Ph.D. chemist then stationed at the Frank J. Seiler Research Laboratory, a basic research lab run by AFOSR at the United States Air Force Academy, who had decided to investigate chemical lasers. McDermott enlisted another Seiler researcher, Major John Viola, previously at the AFWL, to work on the singlet delta oxygen production problem, and the AFOSR initially funded McDermott's work with $10,000. Major Dave Thomas, a physics professor at the United States Air Force Academy, recruited additional personnel from the academy's physics department to further support this effort, including addressing the dilemma of how to accurately measure singlet delta oxygen. During this time, McDermott became convinced from his literature studies and other efforts that a liquid-phase reaction between chlorine and basic hydrogen peroxide could generate sufficient amounts of singlet delta oxygen. Their successes, along with the efforts of Major Ben Loving and Lieutenant Colonel Dana Brabson, convinced the then-head of AFOSR, Dr. Bill Lehmann, to support the growing oxygen iodine laser effort at the Frank J. Seiler Research Laboratory with an additional $50,000, along with another $20,000 contribution from AFWL, all of which provided much-needed

equipment and funded travel that permitted collaboration with other key scientists.[43]

Major McDermott transferred to AFWL in 1977, about the same time that Dr. McKnight departed, and he worked with Major Ron Bousek, who ran the COIL program there. Newly arrived Dr. Dave Benard and his team made a site visit to scientists at Rockwell Science in California, the contractor selected to develop the singlet delta oxygen generator. Rockwell had been experiencing some technical difficulties. Perhaps in haste, Benard took over a Rockwell experiment working with both oxygen and iodine in the contractor's lab, and he determined that an inversion of iodine had in fact occurred. This discovery squelched threats to shut down the program, and AFWL received Rockwell's generator. After briefly heading down what was the wrong path, the team reoriented itself by reexamining earlier work done by academic scientists as well as earlier work done at the Academy. AFWL's Captain Nick Pchelkin continued to improve on McDermott's generator design, significantly increasing the amount of singlet delta oxygen for lasing, while Benard prepared detectors that would record the lasing. Supporting these scientists were AFWL's machine shop, a glass blower, and several Air Force enlisted technicians. Finally, on 30 November 1977, a small spike in light intensity was observed. It was possible that lasing had occurred, but it wasn't until the next day when the experimental team realigned the mirrors that lasing occurred every time the fledgling laser device was turned on.[44]

The laser consisted of four basic parts. First, a generator was required to produce the excited singlet delta oxygen, resulting from a chemical reaction between chlorine gas and basic hydrogen peroxide. Second, a trap removed undesirable chlorine, oxygen, and water vapor that would interfere with the laser's gas kinetics. Third, a spray bar, or nozzle, injected iodine into the oxygen flow the excited oxygen dissociated the iodine transferring energy. Finally, an optical resonator, or mirrors at the ends of a laser cavity, extracted energy from the inverted iodine; in a later version, the residual chlorine and iodine were removed at this stage.[45]

With this exciting advance in 1977, the world's first chemical oxygen iodine laser, COIL, had been invented, yielding a power output of only 4 milliwatts, a mere 0.004 watt. This proof-of-concept demonstration, however, was a continuous wave laser, unlike the pulsed iodine laser invented a decade before, meaning that all the advantages of continuous wave lasers might now be exploited with the use of chemical fuels. Majors McDermott and Bousek, Captain Pchelkin, and Dr. Benard, all of whom were Air Force personnel (three-quarters blue-suiters), authored the Air Force's patent on COIL awarded in 1981.[46]

Once AFWL proved that COIL worked, a race commenced to dramatically increase its power. A second COIL device failed to withstand the required changes and did not lase, but in an almost apocryphal event, Captain Pchelkin and Dr. Benard had already designed the successful COIL III device on the back of a placemat in a local restaurant. By 28 July 1978, less than a year after achieving "first light," its successful first operation, the power jumped to 100 watts in this new version of COIL, an increase nearly 100,000-fold. New collaborators were pulled in, and the

AFWL team tapped the nation's science and technology contractor base to exploit a scheme to generate lasing in a supersonic flow. This effort paid high dividends, as year after year, COIL reached record power levels. By 1982, COIL IV incorporated significant advances and had produced over 4 kilowatts of power. This effort was overtaken in June 1984 when AFWL scientists led by Dr. Gordon Hager created the first supersonic COIL device, called RECOIL, with a power output of 1.5 kilowatts; a second device quickly reached over 4 kilowatts. Also in 1984, Rockwell Corporation (under Air Force contract) developed a supersonic mixing-nozzle array. Supersonic COIL devices, named so because they moved gas supersonically through the cavity, were vastly improved over the subsonic COILs (I through IV). Supersonic devices reduced the size of the laser, reduced the cavity operating temperature, reduced water vapor to increase efficiency, improved the beam quality, and provided greater power output. By 1987, an AFWL COIL device known as ROTOCOIL, using TRW Corporation's rotating disk oxygen generator and Rocketdyne's scalable mixing-nozzle arrays, achieved 25 and then 35 kilowatts of power. Oxygen generator improvements increased efficiency, since COIL efficiency was directly related to the amount of excited oxygen created. Just as important, a COIL device using a high-Q ring resonator produced a near pristine laser-beam quality. These supersonic COIL devices showed that COIL lasers could be scaled to increase their power for weapon applications.[47]

Almost from the beginning, COIL attracted high-level attention. Laboratory commanders and directors, notably Dr. Lehmann (director of AFWL, 19781981) shepherded the program along with generous funding. Air Force Chief of Staff General Lew Allen, a former AFWL officer, visited and encouraged the original COIL device team. When its laser-beam quality reached even higher levels, Secretary of the Air Force E.C. Aldridge, Jr., lauded its potential. Although the Air Force's original intent had been to use COIL as a ground-based laser, General Donald Lamberson, a strong proponent of directed energy technology, quickly became interested in COIL as a potential candidate for the next-generation ABL.[48]

Barely twenty-five years after this groundbreaking discovery by a nearly all-officer team of in-house Air Force scientists, the ABL is on the verge of powering a COIL in the range of several megawatts of power, an increase of nine orders of magnitude roughly a billion times more powerful than the first small laboratory device. As early as 1990–91, when the ABL program began its emergence from its ALL origins, COIL was immediately considered the leading candidate for the ABL's weapon system. Its initial advantages included its continuous short wavelength for better beam propagation through the atmosphere and more energy on target. By the early 1990s, COIL technology had matured to permit a reliable, stable, and safe device offering substantial savings in weight and a relatively minor cost of operation in comparison with other chemical lasers. Its scalability to weapon-system power levels made it the system of choice. When compared to the original and antiquated ALL CO_2 laser, COIL was considerably less bulky and was lighter and able to generate considerably more power, resulting in longer effective ranges.[49]

During the 1990s, significant improvements occurred with COIL. In 1993, power operations were sustained for more than two minutes, and the following year the Phillips Laboratory's VertiCOIL was able to continuously lase for more than ten minutes, and its closed-loop system permitted chemicals to be extracted and reused. Because COIL operates in relatively low temperatures, plastic parts replaced metal ones to increase weight savings and decrease cost. In 1995, Phillips Laboratory scientists built the RADICL, the research assessment device improvement chemical laser. A 10- to 20-kilowatt supersonic COIL, RADICL, tested new laser nozzles designed to permit lower temperatures and improved efficiency. RADICL's laser beam could be transmitted via fiber optics, and a shift from costly helium buffer gas used in the supersonic flow to less expensive nitrogen provided another advancement.[50] In addition, Air Force researchers determined that the high-power laser had significant commercial applications including cutting thick metals under hazardous conditions, for example, in dismantling contaminated nuclear power plants.[51]

AFRL's DE scientists who supported the ABL System Program Office continued to advance the laser to its development as a weapon system. Under the November 1996 ABL contract, the Air Force gave Northrop Grumman the task of developing a megawatt-class COIL weapon. Northrop Grumman had already demonstrated the COIL Baseline Demonstration Laser module the previous August, and it now proceeded to develop the Flight-Weighted Laser Module (FLM), which served as the basic building block for the laser weapon (with several FLMs linked to achieve the necessary power levels). The FLM passed its critical design review in March 1997, and Northrop Grumman proceeded with production. That September, Team ABL proved the critical singlet oxygen generator achieved its necessary production levels, and the FLM reached first light in June 1998. In September 1998, the FLM produced 110 percent of its design power. By March 2002, Northrop Grumman had completed its testing of the first FLM, or LM1, which had reached 118 percent of the laser's design power levels. The contractor proceeded to disassemble the module and send it to Edwards AFB in California for integration into the ABL's flight system. The next step was to build and integrate the final five modules into the demonstrator.[52]

In sum, COIL showed what an Air Force laboratory could do best. Creative Air Force minds had invented the first COIL laser at AFWL. As an in-house program, COIL provided practical experience for more than two decades to Air Force scientists and technicians, both blue-suiters and civilians. The Air Force laboratory system, with additional funding from the AFOSR, had steadily nurtured and developed the technology. As Air Force researchers discovered more innovations and breakthroughs for COIL, new applications for its use were likewise identified. COIL had begun as a candidate for a ground-based laser system, but, because of the inventiveness of laboratory researchers, it evolved as the prime laser for airborne applications, potentially for use in space. On top of these noteworthy events, COIL was even ready to transfer to the civilian commercial market, with potential to revolutionize the industrial use

of lasers. Undeniably, over these more than 25 years, COIL set an impressive standard for military research and development efforts.[53]

AIRBORNE LASER

The development and success of adaptive optics and COIL accounted in large part for the Air Force's decision to move forward with the ABL program. A concerted push for ABL began in 1992 when the Air Force began assessing what type of aircraft was the best aerial platform suited to accommodate a new high-energy ABL. In November 1996, the Air Force awarded a $1.1 billion contract to Boeing, Northrop Grumman, and Lockheed Martin to produce a prototype attack laser aircraft using a commercial 747–400 (freighter) airframe. The purpose of the ABL is to detect, track, and shoot down Scud missiles in their boost stage. This second-generation ABL uses a powerful COIL working in conjunction with a nose-mounted pointer and tracker that incorporates advanced adaptive optics.[54]

The justification and military value of the new ABL is to protect the United States from hostile nations who have in their possession or are in the process of developing ballistic missiles capable of carrying nuclear, biological, or chemical payloads. (More than thirty nations possess ballistic missiles, and the list is growing in an unstable world susceptible to unpredictable terrorist attacks.) Although based in the United States, the ABL will be capable of deploying to any region of the world within twenty-four hours. Once on-station and operating in a standoff mode over friendly territory, the ABL will use three low-power lasers to track enemy missiles from launch through their boost stage. (In its boost phase a missile is most vulnerable because it emits a distinctive infrared signature that is trackable and its travel is at a relatively slow rate of speed along a predicted flight path.) In addition, the three lasers will measure distortions caused by turbulence in the atmosphere between the ABL and its target. That critical information is passed on to the ABL's adaptive optics portion of the fire-control system that directs and sharply focuses a high-power beam (in the megawatt range) on the target at a range of up to several hundred of kilometers.[55]

The beam travels at the speed of light, destroying the missile and dispersing its deadly payload over the enemy's own territory, thereby protecting friendly positions on the ground at the intended target site. Because each missile is destroyed almost as soon as the laser is fired, the ABL operator is afforded the opportunity and flexibility to quickly select and destroy other missile threats in the same area. Another advantage of the ABL is that when it detects a missile launch, it also can pinpoint the location of the launch pad on the ground. That information can then be passed on to tactical fighter aircraft to destroy missile launch sites.[56]

ABL has progressed beyond the conceptual phase and is on the verge of moving into the operational Air Force within the next few years. The Air Force ordered the first ABL aircraft from Boeing in January 1998 to initiate the program's prototype hardware development stage. Two years later, on 22 January 2000, the ABL rolled off the Boeing assembly line and was delivered to the company's aircraft modification facility in Wichita, Kansas. The plane remained there to undergo extensive modifications so it would be able to accept the laser, beam-control system, nose turret, and other components that constitute the total, integrated system. After the modifications were completion in May 2002, the ABL flew to Edwards AFB to begin work on on the integration and installation of all components and subsystems on the aircraft. The ABL is currently going through a comprehensive set of ground tests to validate the design and performance of the entire laser system. Once that phase is completed, ABL flight-testing will take place against a variety of aerial targets. We must keep in mind that the ABL would not be at the point it is today without the laboratory development and technological breakthroughs associated with COIL and adaptive optics over the past forty years.[57]

CONCLUSION: THOUGHTS ON REVOLUTIONARY WEAPONS

Throughout history, epochs and nations have been defined by their ability to visualize the connection between emerging technologies and the security needs of the times. In ancient Greece the phalanx—a close-order fighting formation of heavily armed infantry spearmen troops—with its discipline and speed of maneuver allowed Greek warriors to dominate their part of the world for centuries. Roman Legions—a coherent army combat unit of foot soldiers and cavalry capable of speed and agility—used the Roman short sword as a thrusting weapon at close range, which provided Rome the basis to conquer the world as it was then known. Introduction of the stirrup in the sixth century allowed a horse and rider to go from a means of transportation to a lethal shock weapon, combining the mass and speed of the horse with the thrust of a lance or saber. The armor-mounted knight of Europe was neutralized by the standoff firepower of the long bows of the English yeomen. Over time, long bows and massive fortifications were neutralized by the invention of gunpowder and siege cannons. During the age of discovery, empires were built on the range and firepower of sailing ships, only to be superseded by ships powered by steam. With the invention of rifled barrels and rapid-fire weapons, land warfare later came to be dominated by trench warfare.

In the twentieth century, trench warfare was replaced by maneuver with the adaptation of the internal combustion engine to machines of war. Aircraft emerged to provide the ultimate high ground combined with speed, range, and lethality. Later in the century missiles made it possible to strike any place on the globe within minutes. This same technology made it possible to lift highly effective reconnaissance and communications payloads into space. Combined with air-breathing intelligence, surveillance, and reconnaissance assets, these space assets began to pave the way for a new era in warfare, one in which it would be possible to find, fix, track, target, and engage anything of consequence that moved on the face of the earth or through the atmosphere. Incredibly, all this would be accomplished in near real time!

As we enter the twenty-first century, we find ourselves on the verge of a new breakthrough in warfare with the application of directed energy technology to the battlefield. As advanced sensors and kill mechanisms, directed energy applications in the laser and high-power microwave areas will become the centerpiece of twenty-first century arsenals. We are in an era in which precision and the lack of collateral damage are determinants in the acceptability of weapons. Directed energy weapons with their ability to generate both lethal and nonlethal effects at the speed of light will gain greater acceptance. The nation with the vision to embrace these weapons will dominate the battlefield for the foreseeable future. When combined with near real-time intelligence, surveillance, and reconnaissance assets, the ability to strike quickly with air- and space-based directed energy weapons will revolutionize warfare for surface forces. For an aggressor, sanctuaries will be few and retribution swift when faced with such revolutionary systems as the ABL.

Footnotes and References

Bibliography: Precision Timing, Location, Navigation

Alford, Maj. Dennis L., "History of the Navstar Global Positioning System (1963–1985), Air Command and Staff College Student Report 85–0050.

The Aerospace Corporation (no author), "The Global Positioning System: A Record of Achievement," (1994) (in HQ AFSPC/HO archives).

The Aerospace Corporation (multiple authors), "Satellite Navigation", Crosslink (Summer 2002), Los Angeles, CA

Armatys, M., Masters, D., Komjathy, A., and Axelrad, P., "Exploiting GPS as a New Oceanographic Remote Sensing Tool," Proc. ION-NTM 2000, Anaheim, CA.

Barrows, A., Enge. P., Parkinson, B., and Powell. J., Proc. ION GPS 1995 (September 1995, Palm Springs, CA), pp. 1615–1622.

Bock, Y., S. Wdowinski, P. Fang, J. Zhang, S. Williams, H. Johnson, J. Behr, J. Genrich, J. Dean, M. van Domselaar, D. Agnew, F. Wyatt, K. Stark, B. Oral, K. Hudnut, R. King, T. Herring, S. Dinardo, W. Young, D. Jackson, and W. Gurtner. J. Geophys. Res. 102 (B8): 18013–18033 (1997).

Boutacoff, David A, "Navstar Forecast: Cloudy Now, Clearing Later," Defense Electronics (May 1986).

Bradley, George W. III, "NAVSTAR Global Positioning System Decision", CAMP Military History Symposium (11 May 2001), Rapid City SD.

Bradley, Geroge W. III, "Historical Origins of the Global Positioning System", in *Technology and the Air Force: A Retrospective Assessment* (Washington, D.C.: Air Force History and Museums Program, 1997), The History Conference, Air Force Historical Foundation (24 October 1995), Andrews AFB, MD.

Bundy, Dean "Time, Navigation, and Global Positioning," Space Tracks Bulletin (Nov–Dec 91).

Clarke, Christopher H, ". . . And A Star to Steer By," Defense Electronics (June 1989).

Danchik, R.J. John Hopkins APL Tech. Dig. 19 (1): 18–26 (1998).

Denaro, Robert P. "Navstar: The All-Purpose Satellite," IEEE Spectrum Vol. 18 (May 81).

European Union. Galileo Specification Document [online] 2001. Available: http://europa.eu.int/comm/space/doc_pdf/galileo_431.pdf

Getting, Ivan "The Global Positioning System," IEEE Spectrum (December 1993).

Getting, Ivan and Darrah, John, "Global Positioning, Time Transfer and Mapping," Unpublished Paper, 1996 (in HQ AFSPC/HO archives).

Guier, W.H., and Weiffenbach, G.C. John Hopkins APL Tech. Dig. 18 (2): 178-181 (1997).

Jenkins, J., DCMilitary.com [online] 2001. Available: http://www.dcmilitary.com/navy/tester/7_30/national_news/18414–1.html.

Jennings, C., Alter, K., Barrows, A., Enge, P., and Powell, J., Proceedings of the ION GPS 1999 (September 1999, Nashville, TN,), pp. 1923–1931.

Jennings, C., Charafeddine, M., Powell, J., and Taamallah, S., Proceedings of the 21st Digital Avionics System Conference, (Irvine, CA, 2002, 1B11/1–10).

Katel, P. Wired Magazine [online] 1997. Available: http://www.wired.com/wired/archive/5.07/cemex.html

Misra, P. and Enge, P., Global Positioning System: Signals, Measurements, and Performance. (Ganga-Jamuna Press: Lincoln, MA, 2001).

Parkinson, Bradford W., et al, "Global Positioning System (GPS)", Encyclopedia of Space Science and Technology, Hans Mark, ed. John Wiley and Sons, (2003), Hoboken, NJ.

Parkinson, B., and Axelrad, P., "Closed Loop Orbit Trim Using GPS," 40th Int. Astronaut. Congr. Symp. Astrodynamics (October 1989, Malaga, Spain).

Parkinson, Bradford W. and Spiker, James P. Jr., Axelrod, P. and Enge, P., editors, Global Positioning System: Theory and Applications, Vol I and Vol II, (American Institute of Aeronautics and Astronautics, Inc.: Washington, D.C. 1996).

Parkinson, Bradford.W., et al. "Navigation". Journal of the Institute of Navigation 42 (1): 109–164.

Piscane, V.L. John Hopkins APL Tech. Dig. 19 (1): 4–10 (1998).

Reaser, R., U.S. Air Force Joint Program Office [online] 2002. Available: http://www.ccit.edu.tw/~ccchang/Gps_modernization_ppt.pdf

Roberts, C., Measure & Map 6: 28-31 (2000).

Scheer, Lt Col John F., "Navstar GPS: Past Present and Future," The Navigator, (Winter 1983).

Schlesinger, James R. Chairman, Panel of the National Academy of Public Administration, and Adams, Laurence J. ,Chairman ,Committee of the National Research Council, The Global Positioning System: Charting the Future, for the Congress of the United Sates and the Department of Defense (National Academy Press: Washington, D.C., May 1995).

Smith, Keith, "GPS Devined," The Downlink (March 1966).

Sobel, Dava, Longitude: The True Story of a Lone Genius Who Solved the Greatest Scientific Problem of His Time, (Walker Publishing Co., Inc.: New York, 1995).

Special Issue on Global Navigation Systems. Proc. IEEE 71 (10): (1983).

Swider, R. Department of Defense [online] 2000. Available: http://www.igeb.gov/outreach/iberia-modernization.ppt

U.S. Air Force, Navstar GPS Space Segment/Navigation User Interfaces, [online]. ICD-GPS-200C, 1997. Available: http://gps.losangeles.af.mil/gpsarchives/1000-public/1300-LIB/documents/Other_Data/icdgps200c_irn1thru4.pdf

U.S. Air Force, Factsheet [online] 2001. Available: http://www.af.mil/news/factsheets/JDAM.html

U.S. Coast Guard. Nationwide D[ifferential]GPS Status Report. [online] 2001. Available: http://www.navcen.uscg.gov/dgps/ndgps/default.htm

U.S. DOD and DOT, Federal Radio Navigation Plan, [online] 1999. Available: http://avnwww.jccbi.gov/icasc/PDF/frp1999.pdf

U.S. DOD Joint Program Office, [online] 2002. Available: http://gps.losangeles.af.mil/csel/

U.S. Federal Aviation Administration, Local Area Augmentation System, [online] 2002. Available: http://gps.faa.gov/Programs/LAAS/laas.htm.

U.S. Federal Aviation Administration, Wide Area Aug-mentation System, [online] 2002. Available: http://gps.faa.gov/Programs/WAAS/waas.htm

Notes: Tightening the Circle

1. Polybius, Universal History, www.mcsdrexel.edu/~crorres/Archimedes/Seige/Ploybius.html, pp. 1-4.

2. Robert White, A Foundation for the Future: Celebrating 50 Years of Basic Research, Air Force Office of Scientific Research, Arlington, VA, 2002, pp. 7-27, passim.

3. Hugh McDaid and David Oliver, Smart Weapons , Welcome Rain Publishers, NY, 1997. p. 10.

4. http://www.aerofiles.com/ostfriesland.html; R. Cargill Hall, ed., Case Studies in Strategic Bombardment, Air Force History and Museums Program, Washington,. DC, 1998, p. 13.

5. Case Studies in Strategic Bombardment, p. 13.

6. Ibid.

7. Ibid.

8. Daniel R. Mortensen, "The Air Service in the Great War," in Bernard C. Nalty, Winged Shield, Winged Sword, A History of the United States Air Force, Vol. I, Air Force History and Museums Program, Washington, DC, 1997, p. 68; Donald I. Blackwelder, The Long Road to Desert Storm and Beyond: The Development of Precision Guided Bombs, Air University Press, Maxwell Air Force Base, AL, 1992, p. 4 .

9. Blackwelder, p. 3.

10. David R. Mets, The Quest for a Surgical Strike: The United States Air Force and Laser Guided Bombs, Office of History, Armament Division, Air Force Systems Command, 1987, p. 1.

11. Ibid.

12. David R. Mets, Evolution of Air to Ground Weapons: The USAF Quest for a Precision Strike, 1903-Present, SAAS, undated, briefing in author's files, slides 3,6.

13. Daniel R. Mortensen, "The Air Service in the Great War," in Bernard C. Nalty, Winged Shield, Winged Sword, A History of the United States Air Force, Vol. I, Air Force History and Museums Program, Washington, DC, 1997, p. 69.

14. David R. Mets, The Long Search for a Surgical Strike: Precision Munitions and the Revolution in Military Affairs, Air University Press, Maxwell Air Force Base, AL, 2001, p. 5.

15. Ibid.

16. Ibid.

17. Ibid.

18. Richard P. Hallion, "Precision Guided Munitions and the New Era of Warfare," Air Power Studies Centre Paper Number 53, APSC, RAAF Base, Fairburn ACT,

Australia, ND, p. 3. Available online at: http://www.
fas.org/man/dod-101/sys/smart/docs/paper53.htm.

19. *Ibid.*; Website, *World War II Airpower: Bombs, Weapons, Rockets, Aircraft Ordnance,* http://danshistory.
com/ww2/bombs.shtml., pp. 13-14.

20. Hallion, "Precision Guided Munitions and the New Era of Warfare," p. 4.

21. Dik Alan Daso, *Hap Arnold and the Evolution of American Airpower,* Smithsonian Institution Press, Washington, DC, 2000, pp. 151, 158-160.

22. *Ibid.,* p. 156; Wesley F. Craven and James L. Cate, eds., *The Army Air Forces in World War II,* Vol. 6, Chicago, University of Chicago Press, Chicago, IL 1955, Office of Air Force History reprint, 1985, pp. 178-80.

23. Daso, p. 158.

24. *Ibid.* p. 179.

25. Blackwelder, p. 9.

26. Daso, p. 179. .

27. Daso, pp. 185-86; David R. Mets, *The Quest for a Surgical Strike: The United States Air Force and Laser Guided Bombs,* Office of History, Armament Division, Air Force Systems Command, Eglin AFB, FL. 1987, p. 13.

28. Blackwelder, p. 9.

29. Mets, *The Quest for a Surgical Strike,* p. 13.

30. Blackwelder, p. 10.

31. Hallion, "Precision Guided Munitions and the New Era of Warfare," p. 7.

32. *Ibid.,* p. 14; Website, http://www.vectorsite.net/
twbomb3.html, World War II Glide Bombs, p. 8.

33. Blackwelder, p. 10.

34. *Ibid., pp. 9-10.*

35. Hallion, .p. 9.

36. Mets, *The Quest for a Surgical Strike,* p. 11.

37. Blackwelder, p. 15.

38. Wesley F. Craven and James L. Cate, *Men and Planes,vol. 6 of, The Army Air Forces in World War II,* Office of Aor Force History, Washington, DC., 1955, p. 253, as cited by Blackwelder, p. 15.

39. Blackwelder, p. 15.

40. Website, http://www.vectorsite.net/twbomb3.html, World War II Glide Bombs, p. 7

41. Mets, *Evolution of Air to Ground Weapons: The USAF Quest for a Precision Strike, 1903-Present,* p. 17.

42. Mets, *The Quest for a Surgical Strike,* pp. 15, 21.

43. Mets, *The Quest for a Surgical Strike,* p. 26; Hallion, "Precision Guided Munitions and the New Era of Warfare," p. 8.

44. Mets, *The Quest for a Surgical Strike,* p. 28.

45. *Ibid.,* p. 31.

46. Blackwelder, p. 21.

47. Hallion, "Precision Guided Munitions and the New

48. Mets, *The Quest for a Surgical Strike,* p. 31.

49. *Ibid.,* p. 34.

50. *Ibid.,* p. 35.

51. *Ibid.,* p. 40.

52. *Ibid.,* pp. 43-47.

53. Eduard Mark, *Aerial Interdiction in Three Wars,* Center for Air Force History, Washington, DC, 1994, p. 387.

54. *Ibid*

55. Mets, *The Quest for a Surgical Strike,* pp. 47-48.

56. *Ibid.,* p. 52.

57. *Ibid.,* p.53.

58. *Ibid.,* p. 55; Vernon Loeb, "Bursts of Brilliance," The Washington Post Magazine, December 15, 2002, p. 8.

59. Mets, *The Quest for a Surgical Strike,* p. 56.

60. Vernon Loeb, "Bursts of Brilliance," The Washington Post Magazine, December 15, 2002, p. 8.

61. *Ibid.*

62. Mets, *The Quest for a Surgical Strike,* p. 71.

63. *Ibid., p. 83.*

64. While the U.S. Navy employed the electro-optically guided Walleye at this time, but none of their strike aircraft were certified for LGBs. See: Email, Dave Mets/SAAS to Robert White/AFOSR/PIC, 17 July 2003, subj.: PGM Draft.

65. *Ibid.*

66. Mets, *The Quest for a Surgical Strike,* pp. 86-87; Mark, *Aerial Interdiction in Three Wars,* p. 387; Wayne Thompson, *To Hanoi and Back, The USAF and North Vietnam, 1966-1973,* Air Force History and Museums Program, Washington, DC, 2000, pp. 234-35.

67. Mark, p. 387. Also see: Major Donald L. Ockerman, Tactical Analysis Division, Headquarters, 7th Air Force, Thailand, "An Analysis of Laser Guided Bombs in SEA," 28 June 1973, Air Force Historical Research Agency, K740.041-4.

68. Hallion, "Precision Guided Munitions and the New Era of Warfare," p. 4.

69. Donald I. Blackwelder, *The Long Road to Desert Storm and Beyond: The Development of Precision Guided Bombs,* School of Advanced Airpower Studies, Air University, Maxwell Air Force Base, AL, 1992, p. 30; Michael Rip and James Hasik, *The Precision Revolution: GPS and the Furure of Aerial Warfare,* Naval Institute Press, Annapolis, MD, 2002, p. 224; John A. Tirpak, *Deliberate Force,* Air Force Magazine Online, at: http://
www.afa.org/magazine/oct1997/1097deli.asp, pp. 4-5, citing Air Force Secretary Sheila E. Widnall address to the 1996 AFA Air Warfare Symposium; Kenneth P. Werrell, "Did USAF Technology Fail in Vietnam?" Aerospace Power Journal, Spring 1998, available at:

http://www.airpower.maxwell.af.mil/airchronicles/ apj/apj98/spr98/werrell.html, p. 7; CEP of 447 feet based on F-105 Rolling Thunder campaign employment of unguided munitions and CEP of 23 feet is for F-105 employment of LGB munitions between February 1972 through February 1973; Wayne Thompson, *To Hanoi and Back: The USAF and North Vietnam, 1966-1973*, Air Force History and Museums Program, USAF, Washington, DC, 2000, p. 306. Approximately 8,000,000 tons of air munitions were dropped in SEA by US forces, 6,162,000 figure reflects total tonnage minus B-52 tonnage, both in North and South Vietnam. Also see: *Reaching Globally, Reaching Powerfully: The Uniited States Air Force in the Gulf War, A Report September 1991*, p. 15, at: http://www.fas.org/man/ dod-101/ops/docs/desstorm.htm. A consolidated CEP for the 26,000+ PGMs dropped in Vietnam could not be found, but several periodic CEP measurements exist. Blackwelder notes: "In 1969, 1612 LGBs were dropped with 923 (57%) direct hits, an additional 1114 damaged targets...between 1 February 1972 and 28 February 1973, 10,651 laser guided bombs were dropped in SEA...[n]early 50% (5,107) of the bombs were direct hits...," pp. 26-7. Blackwelder continues: "...during the 1972 Easter Offensive, the Air Force dropped 6000 LGBs...with 64% achieving direct hits." (p. 28). Hallion in *Storm Over Iraq*, notes that "CEPs within six feet [of the target]" were seen during the Linebacker campaign. P. 283. Werrell notes (page 12): "Electro-optical glide bombs (EOGBs) were primarily used in low threat areas. During 1972, 329 were launched, and 53.5% of these achieved direct hits. By comparison, 9,094 LGBs were dropped, and 47.5% achieved direct hits." Blackwalelder (pp. 29-30) also notes: Target acquisition and lock-on with the EO (Walleye), as employed in Vietnam, required a long, stable final approach from 12-14,000 feet AGL at 450 knots. "In low threat areas the crews could make individual passes, but in high threat areas like route Package 6, they had to maintain four-ship pod (to enhance defensive electronic countermeasures) formations...the Walleye was used much less extensively than the laser guided bomb. Weather and restrictive target requirements were the primary limitations. Only 898 Walleye Is, 79 Walleye IIs, and 3 Walleye IIs (with data link) were dropped in SEA through the combined eforts of the Air Force and the Navy...68% of all Walleyes dropped were direct hits." Also see: General Eggers, USAF, "The Employment Effectiveness of Missiles and Guided Weapons," briefing presented to JCS in 1973, Air Force Historical Research Agency, Maxwell AFB, AL, (call no. K712.153-1), p. 56.

70. Hallion, *Storm Over Iraq*, p. 305.

71. *Ibid.*

72. Interview, Robert White with Robert Seirakowski and Al Weimorts, Eglin AFB, FL, 17 June 2003.

73. *Ibid.*

74. *Ibid.*

75. *Ibid*; Email, Dave Mets, SAAS to Robert White, AFOSR/PIC, 17 July 2003, subj.: PGM Draft.

76. E-mail, Martin (Ric) Wehling to Robert White, subj.: Meeting at AFRL/MN Next Week, Atch.: "Tightening the Circle," p. 1.

77. Website article, "Guided Bomb Unit-28 (GBU-28)," www.globalsecurity.org/military/systems/munitions/ gbu-28.htm

78. Website, "The Art of Bunker Busting," http://danshistory.com/lgb.shtml, p. 2.

79. For the most detailed study of the GBU-28 evolution, see: Barry R. Barlow, *The GBU-28, February 1991*, Office of History, Air Force Development Test Center, Eglin Air Force Base, FL, 1996.

80. E-mail, Martin Wehling to Robert White, 18 June 2003, subj: GBU-28 Story.

81. *Ibid.*

82. Eric Schmitt and James Dao, "A Nation Challenged: The Air Campaign, Use of Pinpoint Air Power Comes of Age in New War," New York Times, 24 December, 2001, p. 3.

83. E-mail, Martin (Ric) Wehling to Robert White, subj.: Meeting at AFRL/MN Next Week, Atch.: "Tightening the Circle," p. 2.

84. "A Nation Challenged," pp.3-4.

85. *Ibid., p. 4.*

86. E-mail, Maurice Lourdes, AFRL/AFOSR to Robert White AFRL/AFOSR, 30 August 2003, subj.: FW: S&T Historical Map.

87. INS historical data provided by: Email, AFRL/MN to Robert White/AFRL/AFOSR, subj.: INS Data; 1 August 2003.

88. Michael Rip and James Haslik, *The Precision Revolution*, Naval Institute Press, Annapolis, MD, 2002, p. 224; John A. Tirpak, "Deliberate Force," Air Force Magazine Online, Air Force Association, http://www. afa.org/magazine/oct1997/1097deli.asp

89. See: Adam J. Hebert, "Compressing the Kill Chain," *Air Force Magazine*, March 2003, p. 52.

90. AFOSR/PIP, Briefing, "Anectodal Success Stories," 10 July 2003, slide. 39.

91. Interview, Dr. Robert Sierakowski and Fred Davis, AFRL/MN, 16 June 2003.

92. E-mail, Martin (Ric) Wehling to Robert White, subj.:

Meeting at AFRL/MN Next Week, Atch.: "Tightening the Circle," p. 2; Interview, Dr. Robert Sierakowski and Fred Davis, AFRL/MN, 16 June 2003; John A. Tirpak, "Long Arm of the Air Force," Air Force Magazine, October 2002, p. 32.

93. John A. Tirpak, "Long Arm of the Air Force," Air Force Magazine, October 2002, p. 32.

94. E-mail, Laurice Lourdes, AFRL/AFOSR to Robert White, AFRL/AFOSR, 16 August 2003, subj.: FW: ST Historical Map.ppt.

95. Robert L. Sierakowski, "Meet the Chief Scientist," Technology Horizons, June 2002, p. 12.

96. *Ibid.*

97. Interview, Dr. Robert Sierakowski, AFRL/MN, 16 June 2003.

98. *Ibid.*

99. Martin F. (Ric) Wehling, AFRL/MN, Briefing, "Tara-98: Guidance and Control Technology Overview," 10 February 1998, slides 83-88.

100. AFRL/MN, Briefing, "Micro-Air Delivered Munitions, Advanced Research Program," June 2003, Eglin AFB, FL, passim.

101. Martin F. (Ric) Wehling, AFRL/MN, Briefing, "Tara-98: Guidance and Control Technology Overview," 10 February 1998, slides 89.

102. Interview, Robert White with Robert Seirakowski, Al Weimorts, and Fred Davis, Eglin AFB, FL, 17 June 2003; "Lattice Fin Technology," talking paper, AFRL/MN, undated.

103. See: Buster C. Glosson, "Impact of Precision Weapons on Air Combat Operations," Airpower Journal, Summer 1993, http://www.airpower.au.af.mil/airchronicles/apj/glosson.html, p. 1; Hallion, "Precision Guided Munitions and the New Era of Warfare," passim; Michael Rip and James Haslik, *The Precision Revolution*, passim; Hallion, "Precision Guided Munitions and the New Era of Warfare," passim.

104. Buster C. Glosson, "Impact of Precision Weapons on Air Combat Operations," Airpower Journal, Summer 1993, http://www.airpower.au.af.mil/airchronicles/apj/glosson.html, p. 4.

Footnotes: Enlisting the Spectrum for Air Force Advantage

1. Gen H. H. Arnold, Memo for Dr. Von Karman, 7 November 1994, in Prophecy Fulfilled: "Toward New Horizons" and Its Legacy, M. Gorn ed., Air Force History and Museums Program, Washington, DC, 1994.

2. Ibid.

3. Research sites at Wright-Patterson AFB, OH, also at Hanscom AFB, MA and Kirtland AFB, NM, have been major players in this area.

4. Programs for space-based applications of infrared sensors, not addressed in this paper, fall under Air Force Research Laboratory's Space Vehicles Directorate.

5. U.S. Air Force Fact Sheet, AIM–9 Sidewinder, Public Affairs Office, Air Combat Command, March 2003.

6. James Aldridge, USAF Research and Development: The Legacy of the Wright Laboratory Science and Engineering Community, 1917–1997, Wright-Patterson Air Force Base, Aeronautical Systems Center History Office, OH, 1997, pp.89–96.

7. U.S. Air Force Fact Sheet, LANTIRN, Public Affairs Office, Air Combat Command (as of August 1999); "Martin-Marietta 'LANTIRN' Navigation and Targeting System, Online.. Available: www.wpafb.af.mil/museum

8. Silverman, Mooney, and Shepherd, Infrared Video Cameras, pp. 80–81; Communication from F.D. Shepherd, Aug 03.

9. Silverman, Mooney, and Shepherd, Infrared Video Cameras,

10. R. C. Hardie, K. J. Barnard, J. G. Bognar, E. E. Armstrong, and E. A. Watson, "High-resolution image reconstruction from a sequence of rotated and translated frames and its application to an infrared imaging system," Opt. Eng. 37(1) 247–260, 1998.

11. U.S. Air Force Fact Sheet, B–52 Stratofortress, Public Affairs Office, Air Combat Command (as of July 2001); U.S. Air Force Fact Sheet, Global Hawk, Public Affairs Office, Aeronautical Systems Center, December 2002.

12. D.A. Fulghum, "Advanced Sensors Expand JSF Role," Aviation Week and Space Technology, September 11, 2000, pp.58–59.

13. U.S. Air Force Fact Sheet, RQ–1 Predator UAV, Public Affairs Office, Air Combat Command, May 2002.

14. E. A. Watson, M. P. Dierking, and R. D. Richmond, "Laser Radar Systems for Multi-Dimensional Imaging and Information Gathering," presented at the Annual Meeting of the Lasers and Electro-Optics Society, Orlando, FL, 1998.

15. R. Liebowitz, "Atmospheric Propagation Codes for DoD Systems," ERIM Newsletter, 1990.

16. G. Thompson, "Infrared Target-Scene Simulation Software employed in OIF," Hansconian, June 13, 2003.

17. Silverman, Mooney, and Shepherd, Infrared Video Cameras, p. 82.

18. S.G. Burnay, T.L. Williams, C.H Jones, Applications of Thermal Imaging, Bristol and Philadelphia: Adam

Hilger, 1988, passim.

19. Systems described in this section are from Jane's Electro-Optic Systems, 4th Edition, K. Atkin ed., Butler and Tanner Limited, London, 1999.

20. U.S. Air Force Fact Sheet, AIM–9 Sidewinder.

21. E.H. Tilford, Jr, *USAF Search and Rescue in Southeast Asia*, Washington, DC: Center for Air Force History, 1992, pp. 93, 136.

22. T. Thompson, History of Rome Laboratory, Griffiss AFB, NY: RADC History Office,1997, AF 50th Anniversary Publication, pp. 16–18.

23. History of the Aeronautical Systems Division, January–December 1991, ASC History Office, Wright-Patterson AFB, pp. 336–340.

24. U.S. Air Force Fact Sheet, Global Hawk, Public Affairs Office, Aeronautical Systems Center, December 2002.

25. "Back to Bosnia," *Aviation Week and Space Technology*, October 14, 1996; R.J. Newman, "The Little Predator That Could,"Air Force Magazine, March 2002, pp. 48–53.

26. Maj Gen Stephen G. Wood, Commander, Air Warfare Center, Nellis AFB, Nevada

Footnotes: From the Air

1. Niccolo Machiavelli, *The Prince and the Discourses.* New York: Modern Library, 1950, p. 523.

2. Airborne remote sensing collects and analyzes information of an area or thing from a distance using different kinds of sensors operating from a variety of aircraft.

3. Invented at Stanford University in 1939, the klystron was a high-frequency electron tube. The klystron produced higher power levels than the magnetron, another electron tube used in many types of radar.

4. Maurice Long, ed., Airborne Early Warning System Concepts. Boston: Artech House, 1992, p. 3.

5. Alfred Goldberg, ed., A History of the United Sates Air Force, 1907–1957. Princeton, New Jersey: D. Van Nostrand Company, Inc., p.136.

6. Kenneth Schaffel, *The Emerging Shield: The Air Force and the Evolution of Continental Air Defense.* Washington, D.C., Office of Air Force History, p. 221.

7. Harry A. Pearce."AWACS to Bridge the Technological Gap," Air University Review vol. 23, p. 56, May–June 1972.

8. Lori A. McClelland and David M. Russell, (September 1984). "AEW Market Soars: Interest in Airborne Early Warning Aircraft Runs High Following Combat Demonstrations," Defense Electronics. vol.16, p.127, September 1984.

9. Tom Kaminski and Mel Williams, *The United States Military Aviation Directory*. Norwalk, Connecticut: AIRtime Publishing Inc., 2000, p. 50.

10. "Boeing Company Selected to Build AWACS Aircraft," Department of Defense News Release, July 8, 1970.

11. "Boeing Gets Big Radar Plane Contract," New York Times, July 9, 1970.

12. What is today Air Force Research Laboratory, Information Directorate (AFRL/IF), was known as Rome Air Development Center (RADC) until 1990 and Rome Laboratory until 1997. To avoid confusion, "Rome Laboratory" will be used to designate all research conducted at that location.

13. "AF Demonstrates Pave Mover System," The Hansconian. November 18, 1982.

14. Thomas Robillard, The Joint STARS Platform Decision. Washington, D.C.: Department of the Air Force, 1997, passim.

15. "Pave Mover/Assault Breaker," Rome Research Site History File, n.d.

16. "Joint STARS Flies High in April," Hansconian, June 10, 1988.

17. "Joint STARS Earns Initial Operational Capability," Air Force Materiel Command (AFMC) News Release, December 14, 1999.

18. Kaminski and Williams, The United States Military Aviation Directory, p. 54.

19. James Aldridge, ed., USAF Research & Development: The Legacy of the Wright Laboratory Science and Engineering Community, 1917–1997. Wright-Patterson Air Force Base: Aeronautical Systems Division History Office, n.d., p. 90.

20. "Affordable Moving Surface Target Engagement Enters Second Phase," AFRL News Release, Sep 29, 2000.

21. Rich Tuttle, "Northrop Grumman Beats Ratheon in AMSTE Competition," Aerospace Daily, Sep 20, 2001.

22. Jon Jones, Steven Southwell, Stephen Scott, Stephen Welby, and Charles Taylor,"Affordable Moving Surface Target Engagement," AIAA Missiles and Science Conference, Nov 2002.

23. David Fulghum, "Bull's-Eye Concludes Moving Target Test," Aviation Week & Space Technology. vol 155, pp. 33–34, September 10, 2001

24. Mark Pronobis,"Battlespace Awareness Through Information Fusion," AFRL Technology Horizons. vol. 2, p. 14, Jun 2001.

25. Jim Coghlan, "Countering the Cruise Missile: The Air Defense Initiative Seeks Multispectral Space-based Sensors, Defense Electronics. vol. 19, pp. 77, September 1987.

26. Jeffrey Brandstadt and Mark Kozak, "Multi-Platform Tracking Exploitation," National Symposium and Sensor Fusion Conference, June 2001.

27. Jon Jones and Martin Liggins, "Off-Board Augmented Theater Surveillance," National Symposium and Sensor Fusion Conference, April 1996.

28. "Sensors, Electronics, and Battlespace Environment," Defense Technology Area Plan. Washington D.C.: Department of Defense, 1997, chap. 7, p. 18.

29. Allen A. Boraiko, "The Chip: Electronic Mini-Marvel That Is Changing Your Life," National Geographic. vol. 162, p. 421, October 1982.

30. Aldridge, USAF Research & Development, p. 90.

31. Air Force avionics research at Wright-Patterson Air Force Base is a long one, characterized by frequent organizational changes and redesignations. To avoid confusion, "Avionics Laboratory" is used throughout this paper to identify research carried out at that location.

32. J.F. Rippin, Jr., "Survey of Airborne Phased Array Antennas," Phased Array Antennas, Proceedings of the 1970 Phased Array Antenna Symposium, 2–5 June 1970, Dedham, Massachusetts: Artech House Inc., 1977, pp. 275–87.

33. T.E. Harwell, "Airborne Solid State Radar Technology," E. Brookner, ed., Radar Technology, Dedham, Massachusetts: Artech House Inc., 1977, pp. 275–287.

34. Aldridge, USAF Research & Development, pp. 97–99.

35. H.E. Schrank, "Some Notable Firsts in Array Antenna History," Antenna Applications Symposium, 19–21 September 2001, pp. 231–49.

36. John Clarke, "Airborne Early Warning Radar," Proceedings of the IEEE, vol. 73, pp. 320, February 1985.

37. S. Silver, ed., Microwave Antenna Theory and Design, Vol 12, Radiation Laboratory Series, New York: McGraw Hill Publishing Co., 1949, pp. 286–301.

38. A.F. Kay and A.J. Simmons, "Mutual Coupling of Shunt Slots," IRE Transactions on Antennas and Propagation, vol. 8, pp. 389–400, July 1960.

39. A. Oliner, "The Impedance Properties of Narrow Radiating Slots in the Broadface of Regular Waveguide," Parts I and II, IRE Transactions on Antennas and Propagation, vol. 5, pp. 4–20, January 1957.

40. MITRE and AWACS: A Systems, Engineering Perspective. Bedford, Massachusetts: MITRE Corporation, 1990, passim.

41. "Air Force Completes AWACS Fleet," Air Force Systems Command Newsreview, vol. 28, p. 6, July 1984.

42. McClelland and Russell, "AEW Market Soars," Defense Electronics , p.130.

43. H. Schnitkin, "Unique Joint STARS Phased Array Antenna," Proceedings of the 1989 Antenna Applications Symposium, University of Illinois, 1989.

44. "Early Delivery for Seventh Consecutive Joint STARS," Leading Edge, October 2001.

45. Jon Jones, Steve Scott, and John Cotton, "Advanced Developments for Joint STARS Enhancements," Tri-Service Radar Symposium, April 1997.

46. "Technology: Key to Progress," Commanders Digest, October 30, 1975.

47. "Air Force Gears for AI Buy," Advanced Military Computing. vol. 5, p.3, January 30,1989.

48. Bernard C. Nalty, ed., Wing Shield, Winged Sword: A History of the United States Air Force, Vol II, 1950–1997, Washington D.C.: Air Force History and Museum Program, 1997, p. 287.

49. Jack Robertson, "Viet 'Wall' Will Sense Enemy, Flash Warning to Main HQ," Electronic News, October 30, 1967.

50. Jacob Van Staarveren, Interdiction in Southern Laos, 1960–1968, Washington, D.C.: Center for Air Force History, 1993, p. 271.

51. Edwin Armistead, AWACS and Hawkeyes: The Complete History of Airborne Early Warning Aircraft, St. Paul, Minnesota: MBI Publishing Company, 2002, p. 80.

52. McClelland and Russell, "AEW Market Soars," p. 127.

53. Alexander Levis, AF/ST, to Ruth Liebowitz, Electronic Systems Center/History Office, "Bruce Brown," e-mail, August 5, 2003.

54. Ibid.

55. Ibid.

56. Nalty, Wing Shield, Winged Sword: A History of the United States Air Force, Vol II, 1950–1997, p. 487.

57. Ibid., p. 499.

58. Armistead, AWACS and Hawkeyes, p. 181.

59. Thomas A. Keaney and Eliot A. Cohen, Gulf War Air Power Survey Summary Report, Washington D.C.: Government Printing Office, 1993, p. 109.

60. "Joint STARS Earns Initial Operational Capability," AFMC News Release, February 1998.

61. Benjamin Lanbeth, "Kosovo and the Continuing SEAD Challenge," Aerospace Power Journal, vol.16, p. 14, Summer 2002.

62. David Fulghum, "New Bag of Tricks," Aviation Week & Space Technology, vol. 158, p. 23, 21 April 2003.

63. "RQ–1 Predator Unmanned Aerial Vehicle," News Release Fact Sheet, ACC/PA, May 2002.

64. Nalty, Winged Shield, Winged Sword: A History of the United States Air Force, p.527.

65. Kenneth Munson, ed., Jane's Unmanned Aerial Vehi-

cles and Targets, Alexandria, Virginia: Directory & Database Publishers Assciation, 2001, p. 283.

66. Chuck Roberts, "Operation Iraqi Freedom," Airman, vol. 47, p. 23, May 2003.

67. Fulghum, "New Bag of Tricks," p. 23.

68. David Fulghum, "Battlefield Buzz," Aviation Week & Space Technology, vol. 158, p. 27, March 24, 2003.

69. Maj Gen. Joseph P. Stein, ACC/DO, to Thomas W. Thompson, AFRL/IFOI, "AF Centennial of Flight Paper," e-mail, August 8, 2003.

70. Brig Gen. Kelvin R. Coppock, ACC/IN, "ISR Reachback: An OIF Success Story," Briefing, 19 June 19, 2003.

71. Ibid.

72. Stein to Thompson, e-mail, August 8, 2003.

73. Bistatic radars operated with their transmitters and receivers in different locations. This configuration afforded several advantages, including more angles from which to observe targets, better electronic counter counter measures (ECCM) capability, and advantages attendant with dispersal generally.

References: Pilots in Extreme Environments

1. "Aerospace Medicine", H. G. Armstrong, ed., The Williams and Wilkins Co., Baltimore, 1961.

2. Carey, C. T. "A Brief History of US Military Aviation Oxygen Breathing Systems" http://webs.lanset.com/aeolusaero/Article/Oxygen%Systems%20history — Pt1.htm.

3. "A History of Mitchel Field — The Cradle of Aviation Museum" http://www.cradleofaviation.org/history/airfields/mitchel.html.

4. Payton, G., "Fifty Years of Aerospace Medicine", AFSC Historical Publications, Series No. 67–180, 1968.

5. Dempsey, C. A., "50 Years of Research on Man in Flight", Aerospace Medical Research Laboratory, Published by the United States Air Force, 1985.

6. Heim, J. W., "Aerospace Medicine — The Early Years", unpublished memoirs, 1966.

7. "Oral History Interview: Major General Harry G. Armstrong (Ret), US Air Force Medical Corps, Aerospace Medical Division, Air Force Systems Command, Brooks Air Force Base, United States Air Force, 1981.

8. "Major General Otis O. Benson, Jr. Biography", http://www.Af.mil/bios/bio_4669.shtml, 1981.

9. History of Research in Space Biology and Biodynamics" http://www.hq nasa.gov/office/pao/History/afsp-bio/illustr.htm.

10. Ernsting, J., "Operational and Physiological Requirements for Aircraft Oxygen Systems", AGARD Report No. 697. 7th Advanced Operational Aviation Medicine Course. RAFIAM, Farnborough, Hants, UK. Nov 1983.

11. Ernsting, J. "Minimal Protection for Aircrew Exposed to Altitudes above 50,000 feet". Joint Airworthiness Committee Paper No. 1014, Issue 2, Jan 1983.

12. Sears, William J., "High-Altitude Pressure Protective Equipment: A Historical Perspective", Pages 121–132 in "Raising the Operational Ceiling: A Workshop on the Life Support and Physiological Issues of Flight at 60,000 feet and Above", AL/CF–SR–1995–0021, Edited by Pilmanis, Andrew A. and Sears, William J. December 1995.

13. Webb, J. T. and Pilmanis, A. A, (1995) Altitude decompression sickness: Operational significance. In Pilmanis, A. A. and Sears, W. J, eds. Raising the operational ceiling: A workshop on the life support and physiological issues of flight at 60,000 feet and above. AL/CF–SR–1995–0021. Brooks AFB, TX

14. Sears W. J. "Vaporization of Tissue Fluids at Extreme Altitudes — A Review", SAFE Annual Meetings 1989.

15. King, Kenneth R. and Griswold, Harrison R., "Air Force High Altitude Advanced Flight Suit Study", USAFSAM–TR–83–49 December 1983

16. Byrnes, Donn A. and Hurley, Kenneth D., "Blackbird Rising — Birth of an Aviation Legend", published by Sage Mesa Publications, Los Lunas N.M. 1999

17. Barry, Daniel M., Bassick, John W., " NASA Space Shuttle Advanced Escape Suit Development" SAE Technical Paper #951545, 25th International Conference on Environmental Systems, San Diego, California, July 1995.

18. Shaffstall, Robert M., Morgan, Thomas R. and Travis, Thomas W., "Development of USAF Pressure Breathing Systems", in "Raising the Operational Ceiling: A Workshop on the Life Support and Physiological Issues of Flight at 60,000 feet and Above", AL/CF–SR–1995–0021, Edited by Pilmanis, Andrew A. and Sears, William J. December 1995.

19. Self, B, W. Sears and A. Pilmanis, "Development of the Sustained High Altitude Respiratory Protection and Enhanced Design G Ensemble (SHARP EDGE), AFRL–HE–BR–TR–2001–0175, Feb 2001

Footnotes: Exploiting the High Ground

1. Eileen Shea, compiler, A History Of NOAA, Being A Compilation of Facts And Figures Regarding The Life And Times of The Original Whole Earth Agency, (Washington, D.C.: National Oceanic and Atmospheric Administration, 1987), p. 9, available

on the Internet at http://www.lib.noaa.gov/edocs/noaahistory.html as of 25 June 2003. [Hereinafter cited as Shea, History of NOAA.]

2. National Weather Service Forecast Office Los Angeles/Oxnard, "Cooperative Weather Program," last updated 17 February 2002, available on the Internet at http://www.nwsla.noaa.gov/coop/index.html as of 25 June 2003. This site also describes the prestigious Thomas Jefferson Award, presented annually "to honor cooperative weather observers for unusual and outstanding achievements in the field of meteorological observations. It is the highest award the NWS [National Weather Service] presents to volunteer [weather] observers."

3. Anonymous, "Benjamin Franklin as a Scientist," n.d., available on the Internet at http://sln.fi.edu/franklin/scientist/scientst.html as of 25 June 2003; Keith C. Heidorn, Ph.D., ACM, "Benjamin Franklin: The First American Storm Chaser," 14 July 1998, available on the Internet at http://www.islandnet.com/~see/weather/history/bfrank1.htm as of 25 June 2003. This source marks Franklin as "perhaps America's first meteorological scientist. Others before him had been weather observers, but few set out to observe and then explain weather phenomena." This site also presents Franklin's account of his encounter with the whirlwind. Rita M. Markus, et al., Air Weather Service: Our Heritage, 1937–1987 (Scott Air Force Base, Illinois: USAF Military Airlift Command, 1987), p. 1. [Hereinafter cited as Markus, et al., Air Weather Service.]

4. Shea, History of NOAA, p. 9.

5. Markus, et al., Air Weather Service, p. 1.

6. Markus, et al., Air Weather Service, p. 1.

7. Markus, et al., Air Weather Service, pp. 1–2. John F. Fuller, Thor's Legions: Weather Support to the U.S. Air Force and Army, 1937–1987 (Boston: American Meteorological Society, 1990), pp. 1–7. [Hereinafter cited as Fuller, Thor's Legions.] This source offers greater detail on the early contributions of the U.S. Army in the nation's weather service and notes that "the magnitude of the transfer" of weather-related duties from the Army Signal Corps to the U.S. Weather Bureau "was illustrated by the fact that the Signal Corps budget for meteorology fell from $753,284.70 in 1891 to $31,687.62 in 1892" (p. 7).

8. V. E. Tarrant, Jutland: The German Perspective (Annapolis, MD: Naval Institute Press, 1995), pp. 229–233 covers the role of German airships at the battle.

9. R. Ernest Dupuy and Trevor N. Dupuy, The Encyclopedia of Military History, From 3500 B.C. to the Present, second revised edition (New York: Harper & Row,

Publishers, 1986), p.950.

10. Fuller, Thor's Legions, pp. 13, 15. For more on the weather's impact on American plans and operations see pp. 9–15 of this source.

11. Dr. Helmut E. Landsberg, Geophysics and Warfare [Washington, D.C.: Research and Development Coordinating Committee on General Sciences, Office of the Assistant Secretary of Defense (Research and Development, 1954), p. 4. [Hereinafter cited as Landsberg, Geophysics and Warfare.]

12. Fuller, Thor's Legions, pp. 17–18.

13. Major Scott D. West, USAF, "Warden and the Air Corps Tactical School: Déjà vu?," thesis submitted to the faculty of the School of Advanced Airpower Studies, and published by the Air University Press, Maxwell Air Force Base, AL, October 1999, p. 23.

14. For a recent biography of Mitchell see James J. Cooke, Billy Mitchell (Boulder, CO: Lynne Rienner Publishers, 2002).

15. Jeffrey C. Benton, They Served Here: Thirty-Three Maxwell Men (Montgomery, AL: Air University Press, 1999), pp. 2–3, available on the Internet via the Air University home page at http://www.au.af.mil/au/aul/aupress/Books/Benton/Benton.pdf as of 30 June 2003.

16. Markus, et al., Air Weather Service, p. 4.

17. Fuller, Thor's Legions, p. 20.

18. For details on Appleton's work, see his 12 December 1947 lecture on the ionosphere on the occasion of winning the 1947 Nobel Prize for Physics, available on the Internet at http://www.nobel.se/physics/laureates/1947/appleton-lecture.pdf as of 30 May 2003. For short biographies of Appleton and Kennelly visit the IEEE History Center at http://www.ieee.org/organizations/history_center/legacies/Appleton.html and http://www.ieee.org/organizations/history_center/legacies/kennelly.html, respectively. A short biography of Heaviside can be found on the Internet at http://www-history.mcs.st-and.ac.uk/history/Mathematicians/Heaviside.html.

19. Landsberg, Geophysics and Warfare, pp. 4–5.

20. Landsberg, Geophysics and Warfare, pp. 4, 18; J. M Stagg, Forecast For Overlord: June 6, 1944 (London: Ian Allan Ltd., 1971), p. 5. This source offers a complete account of the meteorology personnel and skills that supported General Dwight D. Eisenhower's decision to postpone the invasion from 5 June to the next day, and "Why then did the weathermen not foresee the rough conditions on the following days and advise General Eisenhower accordingly?" Fuller, Thor's Legions, pp.

35–212 covers meteorological support during World War II on the home front, Europe (including North Africa, the Italian campaign, the strategic bombing effort, and D-Day), the Pacific, and the China, Burma, India theater in some detail.

21. Michael J. Carlowicz and Ramon E. Lopez, Storms From The Sun: The Emerging Science of Space Weather (Washington DC: The Joseph Henry Press, 2002), pp. 114–115.

22. Anonymous, "Boeing B–17 'Flying Fortress,'" available on the Internet at http://www.wpafb.af.mil/museum/air_power/ap16.htm as of 27 June 03; "Anonymous, "Boeing B–29 'Superfortress,'" available on the Internet at http://www.afmuseum.com/ aircraft/b29.html as of 27 June 03.

23. Landsberg, Geophysics and Warfare, pp. 9, 11. For more on Atmospheric Electricity see C. F. Campen, Jr., et al. (eds.), Handbook of Geophysics for Air Force Designers (Hanscom AFB, MA: Air Force Cambridge Research Center, 1957), pp. 9-1 to 9-35.

24. Anonymous, "B–36 Fleet Destroyed by Tornado," updated 14 June 2003, available on the Internet at http://www.cowtown.net/proweb/tornado/tornado.htm as of 2 July 2003.

25. Landsberg, Geophysics and Warfare, pp. 8–13.

26. This equated to being a "5-star general," a rank awarded to only four other American Army leaders during the War: Generals George C. Marshall, Dwight D. Eisenhower, Douglas A. MacArthur, and Omar N. Bradley. According to a brief biography of him at the Arlington National Cemetery web site (at http://www.arlington-cemetery.net/ hharnold.htm) Arnold retired in 1946, but "was honored in 1949 by the rank of General of the Air Force," making him "the first (and only) officer to attain five-star rank in both the United States Army and United States Air Force." According to a U.S. Navy web page (http://www.history.navy.mil/faqs/faq36-1.htm) the U.S. Navy named three Fleet Admirals, or "5-star admirals," during the War: Admirals William D. Leahy, Chester W. Nimitz, and Ernest J. King. Admiral William F. Halsey, Jr., became Admiral of the Fleet in December 1945.

27. Vannevar Bush, Endless Horizons (reprint, NY: Arno Press, 1975) p. iii. For a biography of Vannevar Bush (11 March 1890–30 June 1974) see G. Pascal Zachary, Endless Frontier: Vannevar Bush, Engineer of the American Century (NY: Free Press, 1997). Many credit Bush with establishing the research teaming of the military and academia that ultimately created ARPANET, the precursor to today's Internet. He is also renowned for his visionary article "As We May Think" in the July 1945 issue of Atlantic Monthly, in which "Bush described a futuristic machine called the 'memex' that promised to 'give man access to and command the inherited knowledge of the ages.' He imagined the memex as a work desk with viewing screens, a keyboard and sets of buttons and levers. Printed and written material, even personal notes, would be stored on microfilm, retrieved rapidly and display on screen by a high-speed 'selector.'" (Zachary, Endless Frontier, pp. 261–262.)

28. David N. Spire, Beyond Horizons: A Half Century of Air Force Space Leadership (Washington, DC: Air Force Space Command in Association with Air University Press and available from the U.S. Government Printing Office, 1998) p. 11. For a narrative of the work of two Air Force science and technology visionaries see Dik Daso (Major, USAF), Architects of American Air Supremacy: General Hap Arnold and Dr. Theodore von Kármán (Maxwell Air Force Base, AL: Air University Press, 1997). General Arnold supported Dr. von Kármán's work on the seminal studies on the interrelationships between dominant air power and science and technology, Where We Stand: First Report to General of the Army H. H. Arnold on Long Range Research Problems of the Air Forces with a Review of German Plans and Developments (22 August 1945) and Toward New Horizons (15 December 1945). The first volume of this multivolume publication predicted simply and directly the high-technology skills of the future United States Air Force: Science: The Key to Air Supremacy. On 15 December 1995, the 50th anniversary of the release of Toward New Horizons, the Air Force Scientific Advisory Board released a vision for a potential future for the service, New World Vistas: Air and Space Power for the 21st Century, a multivolume work about the Air Force, technology, and ideas and their roles in providing for the nation's defense.

29. The designation of the Air Force research and development unit at Hanscom AFB changed many times over the years: Cambridge Field Station (1945–1949); Air Force Cambridge Research Laboratories (1949–1951); Air Force Cambridge Research Center (1951–1960); Air Force Cambridge Research Laboratories (1960–1976); Air Force Geophysics Laboratory (1976–1989); Geophysics Laboratory (Air Force Systems Command) (1989–1990); Geophysics Directorate, Air Force Phillips Laboratory (1990–1997); and Battlespace Environment Division, Space Vehicles Directorate, Air Force Research Laboratory (1997–present). For consistency in this paper, references to events between 1945 and 1975 will be listed as AFCRL; between 1976 and

1990 as the Geophysics Laboratory; between 1990 and 1997 as the Geophysics Directorate; and since 1998 to the Battlespace Environment Division.

30. Dr. Ruth P. Liebowitz, "The Air Force's Geophysics Directorate: A 50th Anniversary Retrospective," Eos, Transactions, American Geophysical Union, (Vol. 76, No. 38), September 19, 1995, pp. 371, 381–382. [Hereinafter cited as Liebowitz, "50th Anniversary."] Anonymous, "Alfred Lee Loomis, the legend, ultrasonics and the radar," available on the Internet at http://www.ob-ultrasound.net/loomis.html as of 13 July 2003. Also see Jennet Conant, Tuxedo Park: A Wall Street Tycoon and the Secret Palace of Science That changed the Course of World War II (NY: Simon and Schuster, 2002) for an interesting look at one of America's more unusual laboratories, "hidden in a massive stone castle," where scientists such as Albert Einstein, James Frank, Niels Bohr, and Enrico Fermi could meet.

31. RAND Corporation, Preliminary Design of An Experimental World-Circling Spaceship, Special Anniversary Edition, (Santa Monica, CA: Rand, 1998), inside cover. [Hereinafter cited as RAND, Preliminary Design.]

32. RAND, Preliminary Design, pp. ii, viii, 2.

33. RAND, Preliminary Design, p. 1.

34. William R. Corliss, NASA Sounding Rockets, 1958–1968; A Historical Summary (Washington, DC: Scientific and Technical Information Office, NASA, for sale by the U.S. Government Printing Office, 1971), p. 11–12, also available on the Internet at http://history.nasa.gov/SP–4401/sp4401.htm as of 7 July 2003. This source notes that 300 boxcars full of V–2 parts notwithstanding, "at best only two complete V–2s could be assembled from the original components. Gyros were in especially short supply and 140 more had to be built by American industry." In addition, "many repairs were required" for the parts confiscated in Germany, "and industry had to turn to and manufacture a great variety of missing pieces." This was especially true for the "entirely new scientific 'warheads'" used on the V–2's at WSPG (p. 13). [Hereinafter cited as Corliss, NASA Sounding Rockets.]

35. For a comprehensive overview to the space race see Walter A. McDougall, The Heavens and The Earth: A Political History of the Space Age (NY: Basic Books, 1985). For a description of the V–2 see Gregory P. Kennedy, Vengeance Weapon 2: The V–2 Guided Missile, (Washington DC: Smithsonian Institution Press for the National Air and Space Museum, 1983) and Michael J. Neufeld, The Rocket and the Reich: Peenemünde and the coming of the Ballistic Missile Era (NY: Free Press, 1995). For a look at the competition between the U.S. and Russia for Nazi Germany's rocket scientists see two widely diverse works: Dieter K. Huzel, Peenemünde to Canaveral (Englewood Cliffs, NJ: Prentice-Hall, Inc., 1962, a memoir of working on V–2s in Nazi Germany and escape to American lines and relocation to the U.S. and Linda Hunt, Secret Agenda: The United States Government, Nazi Scientists, and Project Paperclip, 1945 to 1990 NY: St. Martin's Press, 1991), an account by a former Cable News Network investigative reporter.

36. Corliss, NASA Sounding Rockets, p. 13.

37. A. McIntyre, Summary of AFCRL Rocket and Satellite Experiments (1946–1966) (Hanscom AFB, MA: Headquarters Air Force Cambridge Research Laboratories Office of Aerospace Research, AFCRL 66–868, Special Reports, No. 54, December 1966), p. 4; Corliss, NASA Sounding Rockets, p. 15.

38. McIntyre, Summary of AFCRL, pp. 4–7, Corliss, NASA Sounding Rockets, pp. 13, 15. McIntyre lists a total of 67 V–2 launches with entries for 14 AFCRL V–2 launches while Dr. Ruth P. Liebowitz, "Contributions to Space Science and Technology By the Geophysics Directorate, Phillips Laboratory and its Predecessor Organizations," Hanscom AFB, 1992, p.1, in the Phillips Research Site Historical Information Office archives, cites a total of 13 AFCRL V–2 launches. [Hereinafter cited as Liebowitz, "Contributions."]

39. Liebowitz, "Contributions," p. 1; Interview, Dr. Barron K. Oder, Air Force Research Laboratory Space Vehicles Directorate Historical Information Office with Dr. William Denig, Air Force Research Laboratory Space Vehicles Directorate Space Weather Center of Excellence, 20 June 2003.

40. Corliss, NASA Sounding Rockets, p. 15. Note that this source and McIntyre show different dates for the end of V–2 tests; the former indicates on p. 13, "the final V–2 flight on 28 June 1951," while the latter, p. 7, shows an AFCRL V–2 experiment at WSPG on 22 August 1952. A V–2 chronology, located on the Internet at http://www.astronautix.com/lvs/v2.htm as of 7 July 2003 shows its last entry for a V–2 flight on 19 September 1952. Also see Willy Ley, Rockets, Missiles, and Space Travel, revised and enlarged edition (NY: The Viking Press, 1959), pp. 458–459 for another list of V–2 launches. This list, based on General Electric's Project Hermes Final Report, shows a total of 67 V–2 firings in all configurations, with the last coming on 28 June 1951.

41. Lewis Zarem, New Dimensions of Flight, (NY: E.P. Dutton & Co., Inc, 1959), pp. 151–152. According to Eugene M. Emme, Aeronautics and Astronautics: An

American Chronology of Science and Technology in the Exploration of Space 1915–1960 (Washington: NASA, 1961) p. 70, on 26 July 1952, an "Aerobee fired capsule containing two monkeys and two mice to approximately 200,000 feet at Holloman AFB, all recovered unharmed." This followed what Haley, Rocketry and Space Exploration, p. 242, called an "historic flight" on 20 September 1951. Emme, Aeronautics and Astronautics, p. 68 notes on that day "USAF made first successful recovery of animals from a rocket flight when an instrumented monkey and 11 mice survived an Aerobee flight to an altitude of 236,000 feet from Holloman AFB."

42. McIntye, Summary of AFCRL, pp. 5–6; James Van Allen, John W. Townsend, Jr., and Eleanor C. Pressly, "The Aerobee Rocket," in Homer E. Newell, Jr., ed., Sounding Rockets (NY: McGraw-Hill, 1959), pp. 54–70; see especially the Aerobee specifications and sketch on pp. 60–63 and the table summarizing Aerobee firings through 1956 on pp. 64–69. Also see pp. 71–95 for a description of the Aerobee-Hi rocket, and pp. 96–104 and 190–219 for more on the Nike-Deacon and Nike-Cajun rockets.

43. Andrew G. Haley, Rocketry and Space Exploration: The International Story (Princeton, NJ: D. Van Nostrand Company, Inc., 1958) p. 121. For a good background study of how rockets such as the V–2, Aerobee, and others became the instruments of choice for military researchers interested in the upper atmosphere and ionosphere, see David H. DeVorkin, Science With A Vengeance: How the Military Created the US Space Sciences After World War II (NY: Springer-Verlag, 1992).

44. McIntyre, Summary of AFCRL, pp. 5–12; Corliss, NASA Sounding Rockets, pp. 18–21.

45. Liebowitz, "Contributions," p. 1.

46. Captain Robert C. Truax, U.S. Navy, "The Use of Sounding Rockets for Military Research," in Newell, ed., Sounding Rockets, pp. 45–53.

47. Tina D. Thompson, ed., Space Log 1996 (Redondo Beach, CA: TRW Space & Electronics Group, 1997), p. 65. For the impact of Sputnik on Americans see Paul Dickson, Sputnik: The Shock of the Century (NY: Walter & Co., 2001), which weaves personnel recollections of the launch from average individuals into the history of the first years of the space race. Anonymous, "Explorer-I and Jupiter-C: The First United States Satellite and Space Launch Vehicle," available on the Internet at http://history.nasa.gov/ sputnik/expinfo. html as of 9 July 2003. James R. Killian, Jr., Sputnik, Scientists, and Eisenhower: A Memoir of the First Spe-

cial Assistant to the President for Science and Technology (Cambridge, MA: The MIT Press, 1977) provides an "insider's" look at the American response to Sputnik and some of the options facing the nation.

48. Anonymous, "International GeoPhysical Year (IGY)," updated on 11 August 1999, available on the Internet at http://www.hq.nasa.gov/office/pao/History/sputnik/igy.html as of 9 July 2003.

49. Anonymous, "International Geophysical Year," available on the Internet at http:// www.infoplease.com/ce6/sci/A0825345.html as of 9 July 2003. Many books on IGY have appeared over the years, including two references at this site: Sydney Chapman, IGY: Year of Discovery; the story of the International Geophysical Year (Ann Arbor, MI: University of Michigan Press, 1959), and Walter Sullivan, Assault on the Unknown; the International Geophysical Year (NY: McGraw-Hill, 1961). See also Arthur C. Clarke, The Making of a Moon, the Story of the Earth Satellite Program (NY: Harper, 1958) for a contemporaneous look at both the IGY and the first man-made satellites.

50. Walter A. McDougall, The Heavens and the Earth (NY: Basic Books, 1985), p. 168. This page also recounts a central irony of the early space race: the Army and Navy were in serious competition with each other as much as the Soviet Union for the prestige associated with the first space flight. An added irony came later, when the Air Force became the service most closely associated with space technologies and breakthroughs.

51. Emme, Aeronautics and Astronautics, p. 98, also contains quote from Van Allen.

52. Dr. James L. Green, "The Magnetosphere," n.d., available on the Internet at http:// ssdoo.gsfc.nasa.gov/education/lectures/magnetosphere.html as of 12 July 2003.

53. Dr. David P. Stern and Dr. Mauricio Peredo, "Some Dates in the Exploration of the Magnetosphere," updated 25 November 2001, available on the Internet at http://www-spof.gsfc.nasa.gov/Education/whchron.html as of 12 July 2003.

54. Dr. David P. Stern and Dr. Mauricio Peredo, "#12a. The Inner Radiation Belt," updated 25 November 2001, available on the Internet at http:// www-spof.gsfc.nasa.gov/ Education/winbelt.html as of 12 July 2003; Dr. David P. Stern and Dr. Mauricio Peredo, "#12b. The Outer Radiation Belt," updated 25 November 2001, available on the Internet at http:// www-spof.gsfc.nasa.gov/ Education/woutbelt.html as of 12 July 2003.

55. William E. Burrows, This New Ocean: The Story of the First Space Age (NY: Modern Library, 1998) pp. 211–212 [Hereinafter cited as Burrows, New Ocean];

Michael J. Carlowicz and Ramon E. Lopez, Storms From The Sun: The Emerging Science of Space Weather (Washington DC: The Joseph Henry Press, 2002), pp. 66–71. [Hereinafter cited as Carlowicz and Lopez, Storms.] David P. Stern, "A Brief History of Magnetospheric Physics During the Space Age," last updated on 25 November 2001, available on the Internet at http://www-spof.gsfc.nasa.gov/Education/bh2_1.html. Note that this is an electronic of the original article that appeared in Reviews of Geophysics, 34 (1996) pp. 1–31.

56. Burrows, New Ocean, p. 212, which also notes that Van Allen "would be awarded the Crafoord Prize, comparable to the Nobel Prize, by the Royal Swedish Academy of Sciences in 1989."

57. Philip Taubman, Secret Empire: Eisenhower, The CIA, and the Hidden Story of America's Space Espionage (NY: Simon and Schuster, 2003) pp. 3–34 provides a good overview to President Eisenhower's conundrum about authorizing spy flights and the need to know what "the other side" was up to in terms of military readiness and basing.

58. Spires, Beyond Horizons, pp. 51–53; Burrows, New Ocean, pp. 166–170 provides some of the behind-the-scenes planning by U.S. advisors and lawyers prior to the launch of any satellite as a means of establishing America's right to fly reconnaissance satellites for intelligence purposes.

59. Killian, Sputnik, Scientists, and Eisenhower, pp. 119–144 provides an inside look at how President Eisenhower and his advisors shaped the creation and mission of NASA, and the Department of Defense's position that it must be able to maintain "the latitude to pursue those things that are clearly associated with defense objectives." (p. 136).

60. According to T. M. Pearce (ed.) New Mexico Place Names: A Geographical Dictionary (Albuquerque: University of New Mexico Press, 1990), p. 161, Sunspot is 18 miles south of Cloudcroft, and was so named because of the observatory's mission "to predict sun-induced disturbances in earth's atmosphere and in outer space."

61. Liebowitz, "50th Anniversary," p. 381.

62. Dr. Ruth P. Liebowitz, "Air Force Geophysics 1945–1995: Contributions to Defense and to the Nation," April 1996, pp. 39–40, copy in the archives of the Phillips Research Site Historical Information Office, Kirtland AFB, NM. [Hereinafter cited as Liebowitz, "Air Force Geophysics."]

63. Dr. David H. Hathaway, "Solar Flares," 6 January 03, available on the Internet at http:// science.msfc.nasa. gov/ssl/pad/solar/flares.htm as of 14 July 2003.

64. Anonymous, "Solar flares produce quakes deep inside the sun," 27 May 1998, available on the Internet at http://www.cnn.com/TECH/space/9805/27/sun. quakes as of 14 July 2003.

65. Anonymous, "Thanksgiving Solar Flares Trigger Large Weekend Space Storm," 25 November 2000, available on the Internet at http://www.space.com/spacewatch/space_ weather_001125.html as of 14 July 2003.

66. Dr. Donald Neidig, "New Optical System Automatically Monitors Solar Activity," in AFRL Technology Horizons, June 2001, pp. 29–30, also available on the Internet at http://www.afrlhorizons.com/ Briefs/June01/VS0009.html as of 14 July 2003.

67. Liebowitz, "Air Force Geophysics 1945–1995," p. 37.

68. Liebowitz, "Air Force Geophysics," pp. 41, 48; Liebowitz, "Contributions," p. 5.

69. Liebowitz, "Contributions," p. 5.

70. McDougall, The Heavens and the Earth, pp. 344–360, in a chapter cal "Benign Hypocrisy: American Space Diplomacy" provides a good overview to the evolution of U.S. military and civilian paths in space science; quotes from sidebar on p. 353. Spires, Beyond Horizons, pp. 53–80 provides another look at the development of military and civilian roles in space.

71. Spires, Beyond Horizons, pp. 86–112.

72. James M. Grimwood, Project Mercury: A Chronology, NASA Special Publication 4001, available on the Internet at http://history.nasa.gov/SP-4001/contents. htm as of 11 July 03. This source also indicates that on 7 September "The results of a joint study by the Atomic Energy Commission, the Department of Defense, and NASA concerning the possible harmful effects of the artificial radiation belt created by Operation Dominic on Project Mercury's flight MA–8 were announced. The study predicted that radiation on outside of capsule during astronaut Walter M. Schirra's six-orbit flight would be about 500 roentgens but that shielding, vehicle structures, and flight suit would reduce this dosage down to about 8 roentgens on the astronaut's skin. This exposure, well below the tolerance limits previously established, would not necessitate any change of plans for the MA–8 flight." The July nuclear test was Operation Dominic.

73. Anonymous, "40 Years of Telstar," n.d., available on the Internet at http://www. lucent.com/minds/telstar/fit. html as of 11 July 2003.

74. McDougall, Heavens and the Earth, p. 358.

75. Anonymous, "The Air Force Weapons Laboratory's Measurements of the Space Environment," n.d., in the

archives of the Air Force Research Laboratory Phillips Research Site Historical Information Office, 11 July 2003.

76. Dr. Ruth P. Liebowitz, Chronology: From the Cambridge Field Station to the Air Force Geophysics Laboratory, 1945–1985 (Hanscom AFB: Air Force Geophysics Laboratory, 1985), p. 66.

77. Liebowitz, "50th Anniversary," p. 371.

78. Thomas D. Damon, Introduction to Space: The Science of Spaceflight, 2nd ed., (Malabar, FL: Krieger Publishing Company, 1995), p. 60.

79. Carlowicz and Lopez, Storms, p. 75.

80. Carlowicz and Lopez, Storms, pp. 129–130; and Chris Tschan (USAF, Ret), as quoted on p. 131.

81. John W. Freeman, Storms in Space (Cambridge, UK: Cambridge University Press, 2001), pp. 14–15. [Hereinafter cited as Freeman, Storms.]

82. Jurgen Buchau, et al., "The Digital Ionospheric Sounding System Network of the US Air Force Air Weather Service," presented at XXIIVth General Assembly of the International Union of Radio Science, Kyoto, Japan, 25 August – 2 September 1993, available on the Internet at http://www.ips.gov.au/IPSHosted/INAG/uag-104/index.html as of 14 July 2003.

83. Interview, Oder with Dr. Denig, 20 June 2003.

84. Interview Oder with Dr. Denig, 20 June 2003; Marcio H. O. Aquino, et al., "GPS Based Ionospheric Scintillation Monitoring," available on the Internet at http://www.estec.esa.nl/wmwww/wma/spweather/workshops/SPW_W3/PROCEEDINGS_W3/Aquino.pdf as of 14 July 2003.

85. Freeman, Storms, pp. 61–62.

86. Freeman, Storms, pp. 63–64. See also J. Freeman, et al., "The Magnetospheric Specification and Forecast Model," available on the Internet at http://hydra.rice.edu/ ~freeman/ding/www/ msfm95/msfm.html as of 15 July 2003 provides a thorough background and description of the forecasting system, as well as details on many aspects of the model.

87. Anonymous, "Point Paper: The Geophysics Directorate," Solar News, July 1994, available on the Internet at http://gong.nso.edu/SolarNews/07_94.html as of 12 July 2003.

88. Interview, Oder with Dr. Denig, 20 June 2003.

89. Carlowicz and Lopez, Storms from the Sun, p. x.

90. Anonymous, "Solar Wind Interplanatary [sic] Measurements (SWIM) Help Forecast Space Weather," in Phillips Laboratory Success Stories 1995 (Kirtland AFB, NM: Phillips Laboratory, 1996), p. 71; Air Force Research Laboratory Space Vehicles Directorate Fact Sheet, "Solar Wind Interplanetary Measurements," current as of January 1998, available on the Internet at http://www.vs.afrl.af.mil/Factsheets/swim.html as of 13 July 2003.

91. Anonymous, "Mitigation of Spacecraft Charging," in Phillips Laboratory Success Stories 1995 (Kirtland AFB, NM: Phillips Laboratory, 1996), p. 33; Dr. Keith E. Holbert, "Spacecraft Charging," updated 20 February 2003, available on the Internet at http://www.eas.asu.edu/~holbert/eee460/spc-chrg.html as of 13 July 2003. This last reference provides a link to an informative spacecraft charging flash animation.

92. Alice B. McGinty (ed.), "Air Force Geophysics Laboratory Report on Research For the Period January 1979–December 1980," April 1982, AFGL–TR–82–0132, pp. 101–102; Mark Wade, "SCATHA," updated on 26 June 2002, available on the Internet at http://www.astronautix.com/craft/scatha.htm as of 13 July 2003. Also see data on SCATHA's payload at the Jet Propulsion Laboratory Mission and Spacecraft Library at http://msl.jpl.nasa.gov/QuickLooks/scathaQL.html available at of 13 July 2003. Alice B. McGinty (ed.), "Air Force Geophysics Laboratory Report on Research For the Period January 1981–December 1982," August 1983, AFGL–TR–83–0198, p. 83

93. Charles C. Mann, "The End of Moore's Law?", n.d., available on the Internet at http://www.uow.edu.au/~hasan/buss951/The_future_of_Moores_Law.pdf as of 15 July 2003.

94. Liebowitz, "Air Force Geophysics," p. 33; M. H. Johnson and John Kierein Ball, "Combined Release and Radiation Effects Satellite (CRRES): Spacecraft and Mission," available on the Internet at http://tide1.space.swri.edu/CRRESpaper.html as of 15 July 2003. For the full story on SPACERAD and CRRES see Ruth P. Liebowitz, "A History of the Space Radiation Effects (SPACERAD) Program for the Joint USAF/NASA CREES Mission, Part I: From the Origins Through the Launch, 1981–1990," 16 March 1992, Phillips Laboratory Technical Report PL–TR–92–2071, Special Reports, No. 269.

95. Anonymous, "Mitigation of Spacecraft Charging," in Phillips Laboratory Success Stories 1995 (Kirtland AFB, NM: Phillips Laboratory, 1996), p. 33.

96. U.S. Air Force Fact Sheet, "Scintillation Network Decision Aid," current as of June 2002, available on the Internet at http://www.vs.afrl.af.mil/Factsheets/scinda.html as of 14 July 2003.

97. U.S. Air Force Fact Sheet, "Communication/Navigation Outage Forecasting System," current as of April 2002, available on the Internet at http://www.vs.afrl.af.mil/ Factsheets/cnofs.html as of 14 July 2003.

98. Carlowicz and Lopez, Storms from the Sun, pp. 13–16.

99. Air Force Research Laboratory Space Vehicles Directorate Fact Sheet, "Solar Mass Ejection Imager," current as of November 2002, available on the Internet at http://www. vs.afrl.af.mil/Factsheets/SMEI.html as of 13 July 2003; Lockheed Martin Corporation Press Release, "Titan II Successfully Launches Coriolis Mission from Vandenberg Air Force Base," 6 January 2003, available on the Internet at http://www.spaceref.ca/news/ viewpr.html?pid=10319 as of 15 July 2003.

100. E-mail message, Janet C. Johnston, Air Force Research Laboratory Space Vehicles Directorate Battlespace Environment Division to Barron K. Oder, AFRL Space Vehicles Directorate Historical Information Office, "RE: History Tasker," 6 February 2003.

101. Anonymous, "Timeline of the Twentieth Century: 1900–1909, n.d., available on the Internet at http:// history1900s.about.com/library/time/bltime1900. htm; Tim Dirks, "The Great Train Robbery," available on the Internet at http://www.filmsite.org/grea. html; Anonymous, "Harley-Davidson Motorcycles: The American Legend," n.d., available on the Internet at http://www.carpartsplaza.com/harley_story/htm; Anonymous, "From a Dream to a Legacy," n.d., available on the Internet at http://www.spartanburghog. com/ hdhistory html; and Anonymous, "Ford Motor Company history timeline," n.d., available on the Internet at http://wwwautomuseum.com/forddate. html, all available as of 14 July 2003.

102. Air Force Space and Missile Systems Center Fact Sheet, "Defense Meteorological Satellite Program, " current as of June 2003, available on the Internet at http://www. losangeles.af.mil/SMC/PA/Fact_Sheets/ dmsp_fs.htm as of 15 July 2003; Anonymous, "DMSP transferring control of satellites to NOAA," 15 July 1998, available on the Internet at http://www.af.mil/ revised/n19980715_981032.txt as of 23 May 2003; Neal Peck, "Historical S&T," n.d. [Dec 99], in the archives of the Phillips Research Site Historical Information Office as of 15 July 2003. For an in-depth look at the Air Force Geophysics Laboratory's role in DMSP see Dr. Ruth P. Liebowitz, "Historical Brief: AFGL Support to the Defense Meteorological Satellite Program, 1971–1988," August 1988, in the archives of the Phillips Research Site Historical Information Office as of 15 July 2003.

Bibliography: Military Satellite Communications

Armacost, Michael H. The Politics of Weapons Innovation: The Thor-Jupiter Controversy. New York: Columbia University Press, 1969.

Beard, Edmund. Developing the ICBM: A Study in Bureaucratic Politics. New York: Columbia University Press, 1976.

Chapman, John L. Atlas: The Story of a Missile. New York: Harper & Brothers, 1960.

Dyson, George. Project Orion: The True Story of the Atomic Spaceship. New York: Henry Holt, 2002.

Hallion, Richard P., ed. The Hypersonic Revolution: Case Studies in the History of Hypersonic Flight. 3 Vols. Bolling AFB, DC: Air Force History and Museums Program, 1998.

Hart, Julian. The Mighty Thor: Missile in Readiness. New York: Duell, 1961.

Heppenheimer, T.A. The Space Shuttle Decision: NASA's Search for a Reusable Space Vehicle. Washington, DC: NASA History Office, 1999.

Launius, Roger D., and Dennis R. Jenkins, eds. To Reach the High Frontier: A History of U.S. Launch Vehicles. Lexington: University of Kentucky Press, 2002.

MacKenzie, Donald A. Inventing Accuracy: A Historical Sociology of Nuclear Missile Guidance. Cambridge: MIT Press, 1990.

Miller, Jay. The X–Planes: X–1 to X–45. 3rd ed. Hinckley, United Kingdom: Midland Counties Publications, 2001.

Neufeld, Jacob. The Development of Ballistic Missiles in the United States Air Force, 1945–1960. Washington, DC: Office of Air Force History, 1990.

Reed, Dale R., with Darlene Lister. Wingless Flight: The Lifting Body Story. Washington, DC: NASA History Office, 1999.

Stumpf, David K. Titan II: A History of a Cold War Missile Program. Fayetteville: University of Arkansas Press, 2000.

Thompson, Milton O. At the Edge of Space: The X–15 Flight Program. Washington, DC: Smithsonian Institution Press, 1992.

Thompson, Milton O., and Curtis Peebles. Flying Without Wings: NASA Lifting Bodies and the Birth of the Space Shuttle. Washington, DC: Smithsonian Institution Press, 1999.

Footnotes: Directed Energy

1. Greg Caires, "Fogleman: Energy Weapons Will Be Centerpiece Of Future AF," Defense Daily, July 23, 1998, p. 3.

2. Interview, Robert W. Duffner with Dr. Petras Avizonis, December 15, 1988; Interview, Duffner with Major General Donald L. Lamberson, January 12 1989.

3. Interview, Duffner with Dr. Hans Mark (former Secretary of the Air Force), September 29, 1992.

4. T.H. Maiman, "Stimulated Optical Radiation in Ruby," Nature, August 6, 1960, pp. 493–494; John A. Osmundsen, "Light Amplification Claimed by Scientist," New York Times, July 8, 1960, pp. 1, 7.

5. Arthur L. Schawlow, "Lasers: The Practical and the Possible," The Stanford Magazine, Spring/Summer, 1979, pp.24–29.

6. Interview, Duffner with Lamberson, January 12, 1989; Jeff Hetch, Beam Weapons: The Next Arms Race, New York: Plenum Press, 1984, pp. 1–13.

7. Rpt, "Laser Research and Applications," prepared at the request of the Honorable Howard W. Cannon, Chairman, Committee on Commerce, Science, and Transportation, U.S. Senate, Washington, D.C., U.S. Government Printing Office, 1980; Rpt, Colonel John C. Scholtz, Jr., "The Air Force High-Energy Laser Program," Air War College Professional Study No. 4229, November 1970, pp. 1–6.

8. Interview, Duffner with Demos T. Kyrazis, (Director of ALL), August 8, 1985; Phillip J. Klass, "Special Report: Laser Weapons," Aviation Week & Space Technology, August 18, 1975, pp. 34–39.

9. Statement by Hans Mark, Secretary of the Air Force, Before the Subcommittee on Science, Technology, and Space, Albuquerque, New Mexico, January 12, 1980; Rpt, U.S. Senate, "Laser Research and Applications," 1980.

10. Ibid.

11. Interview, Duffner with Kyrazis, 8 August 1985; Interview, Duffner with Russ Parsons (Colonel, USAF, Retired), December 15, 1988; Anthony Ripley, "Laser is Stirring Imagination of Weapon Scientists," New York Times, October 16, 1971, p. 15.

12. George Heilmeir, "Contributions to Science and Technology by the Advanced Research Projects Agency (ARPA) 1957–1970," Commanders Digest, October 7, 1976; Interview, Duffner with David R. Jones (former AFWL Commander), September 19, 1988.

13. Interview, Duffner with Jones, September 19, 1988; Ltr, J.P. Ruina, Director ARPA, to Commander, Office of Aerospace Research, "ARPA Order 313–62," February 26, 1962.

14. Interview, Duffner with Lamberson, January 12, 1989; Rpt, RTD-TDR-63-3070, Marvin C. Atkins et al. "AFWL Progress on PROJECT SEASIDE as of July 20, 1963," November 1963, pp. iii, 1, 41–44.

15. Interview, Duffner with Lamberson, January 12, 1989; C. Kumar N. Patel, "Selective Excitation Through Vibrational Energy Transfer and Optical Laser Action in N2-CO2," Physical Review Letters, Vol 13, 1964, pp. 617–619.

16. Paper, Air Force Weapons Laboratory, "A Thumbnail Sketch of the US HEL Weapons Program," n.d., p. 1; Rpt, WLREL–69–025, "Advanced Development Program: High-Energy Laser Program, Program 644A–Task II," February 24, 1969, pp. A1–A3.

17. Letter Contract, F29601–69–C–0058, Air Force Special Weapons Center with Hughes Aircraft Company for "Optics For High Power Systems," February 3, 1969; Rpt, AFWL, Colonel David R. Jones, "The Air Force Weapons Laboratory Management Plan: EIGHTH CARD Program," February 13, 1969, p. 5.

18. Interview, Duffner with Kyrazis, June 20, 1985.

19. Rpt, ARTO–73–1, Lamberson, "Program Management Plan: Advanced Radiation Technology Office," June 1973, pp. 2–3; Briefing Chart, AFWL/ARTO, "TSL Milestones," September 7, 1972.

20. Ltr, Colonel Allison, Chief, Office of Security Review, Office of Public Affairs, HQ USAF, to AFSC/PAS, "Fact Sheet and Film — DELTA," August 29, 1980.

21. Interview, Duffner with Lamberson, January 12, 1989; Rpt, AFWL–TR–86–01, Vol I, Raymond V. Wick, "Airborne Laser Laboratory Cycle III: Systems and Test Descriptions," May 1988, p. v.

22. Briefing, Lamberson, "Milestone Review of Cycle II," June 8,1976

23. Rpt, AFWL, "Airborne Laser Laboratory Cycle III Test Plan (Executive)," October 31, 1977, pp. 3–5, 93–94; John C. Rich, "Air Force Overview," in Proceedings of the Third DoD High-Energy Laser Conference, July 1979, pp. 47, 52–53; John W. Dettmer and Demos T. Kyrazis, "Overview of the Airborne Laser Program," in Proceedings of the Third DoD High-Energy Laser Conference, July 1979, pp. 379–396.

24. Interview, Duffner with Colonel John Otten (former Test Director ALL), May 26, 1994; Interview, Duffner with Colonel Jerry Janicke (former Chief, AFWL's Laser Development Division), October 4, 1988; Ltr, Major General Jasper A. Welch, Jr., Asst DCS/Research, Development and Acquisition, HQ USAF, to AF/CC et al., "Item of Interest," May 26, 1983.

25. Interview, Duffner with Lieutenant Colonel Denny Boesen (former Test Director for ALL), August 12, 1994; Interview, Duffner with Lieutenant Colonel Steve Coulombe, February 6, 1995; Interview, Duffner with General Robert T. Marsh (former Air Force Systems Command Commander), August 13, 1990; Rpt, AFWL–TR–86–01, p. 94; USAF News Release (AFSC #83–130), November 29,1993; Video, "Airborne Laser Laboratory Laser Flight," March 1, 1985.

26. H. W. Babcock, "The possibility of compensating astronomical seeing,": Publ. Astron. Soc. Pac., Vol. 65, pp 229–236, 1953.

27. R. B. Leighton, "Concerning the problem of making sharper photographs of the planets," Scientific American. vol. 194, June 1956, pp 157–166,

28. J. W. Hardy, J. Feinlieb, and J. C. Wyant, "Real-time phase correction of optical imaging systems," Dig. Technical Reports, Topical Meeting on Optical Propagation Through Turbulence, sponsored by OSA, Boulder, CO, July 1974.

29. A. Buffington, F. S. Crawford, R. A. Muller, and C. D. Orth, "First observatory results with an image sharpening system," J. Opt. Soc. Amer., Vol. 67, 1977. pp. 304–305.

30. Charles D. Corsetti, Stephen B. Talley, and Charles W. Martin, "Design of a digital control system for an advanced optical tracker," Midwest Symposium on Circuits and Systems, June 1981.

31. John W. Hardy, "Adaptive Optics," Scientific American, June 1994, pp 60–65.

32. R. Q. Fugate, et al, "Two generations of laser-guide-star adaptive-optics experiments at the Starfire Optical Range," JOSA, Vol. 11, No. 1, 1994, pp. 310–324.

33. Robert R. Beland, "Propagation through Atmospheric Optical Turbulence," The Infrared & Electro-Optical Systems Handbook, Vol. 2, Chapter 2, SPIE Press, Bellingham WA, 1993.

34. Ronald J Hugo, et al, "Acoustic noise-source identification in aircraft-based atmospheric temperature measurements," AIAA Journal, Vol. 40, No. 7, July 2002.

35. AFRL–DE–TR–2002–1024, Lawrence D. Weaver, et al., "ABLSTAR," 2002.

36. Russell Butts and Lawrence D. Weaver, "ABLEX: High Altitude Laser Propagation Experiment," Proceeding of the NASA–UCLA Workshop on Laser Propagation in Atmospheric Turbulence, edited by A. V. Balakrishnan and Russell Butts, February 1994.

37. D. C. Washburn, et al, "Airborne Laser Extended Atmospheric Characterization Experiment (ABLE ACE)," PL–TR–96–1084, 1996.

38. Rpt, AFWL/AR, "Advanced Radiation Technology: PE 63605F," September 1985.

39. Glen P. Perram, Chemical Lasers, Air Force Institute of Technology; "What's a Chemical Laser?" [Online] Available: http://www.aae.uiuc.edu/~detweile/chemlase.htm; Rpt, William McDermott, A History of the Oxygen Iodine Chemical Laser, n.d., 1979, pp. 5, 7.

40. McDermott, A History, pp. 1–2.

41. Ibid., pp. 2–3; William E. McDermott, "Historical Perspective of COIL" in Steven J. Davis and Michael C. Heaven, editors, Gas and Chemical Lasers and Intensive Beam Applications III, SPIE Proceedings Volume 4631, 2002, p.1.

42. McDermott, A History, pp. 3–4; McDermott, "Historical Perspective," p.2, 5.

43. McDermott, A History, pp. 10–14; McDermott, "Historical Perspective," p.2, 3.

44. McDermott, A History, pp.17–25; McDermott, "Historical Perspective," pp.4–5; Interview, Duffner with Nicholas Pchelkin, August 24, 1999; Ltr, Edward G. Clements, Tech Advisor, Programs and Requirements Office, to AFSC/DLZ, subj: Outstanding Technical Achievements, January 3, 1978; W. E. McDermott, et al., "An electronic transition chemical laser," Applied Physics Letters, April 15, 1978, pp. 469–470.

45. Rpt, AFWL Oxygen-iodine Chemical Laser, n.d.; Briefing to AFSC/XTW, Lt Col Hasen, AFWL/ARDI, "Chemical Oxygen Iodine Lasers (COIL), May 17, 1989; AFRL/DE Fact Sheet, "Chemical Oxygen Iodine Laser," January 1998.

46. McDermott, A History, p.25; Interview, Duffner with Pchelkin; United States Patent, McDermott et al, Continuous Wave Chemically Pumped Atomic Iodine Laser (No. 4,267,526), May 12, 1981; United States Patent, McDermott et al., Gas Generating System for Chemical Lasers (No. 4,246,252), January 20, 1981.

47. McDermott, A History, pp.25–29; McDermott, "Historical Perspective," pp. 5–6; K.A. Truesdell, C.A. Helms, and G.D. Hager, "COIL Development in the USA," AIAA Paper 94–2421 presented at the 25th AIAA Plasmadynamics and Lasers Conference, Colorado Springs, CO, June 1994, p.5; Ltr, AFWL/CC to AFSC/DL, "COIL Beam Quality Milestone," July 23, 1987; AFWL TR–85–43, Vol I, J.O. Berg, et al., "Oxygen Iodine Supersonic Technology," October 1985, pp. 35–40; Ltr, Maj Gerald Hasen, AFWL/ARDA to AFWL/AR, "Nomination for AFWL Technical Achievement Award," January 30, 1985.

48. Ltr, Secretary of the Air Force E. C. Aldridge, Jr., to General Bernard P. Randolph, AFSC/CC, subj: COIL, August 5, 1987.

49. McDermott, "Historical Perspective," Briefing, Colonel John Otten, PL/LI, "Airborne Laser," April 27, 1992; Ltr, PL/SX to SMC/CC. et al., "Airborne Laser Program" October 22, 1992; "Airborne Laser: History" [Online]. Available: http://www.airbornelaser.com/history, July 25, 2003.

50. Air Force Research Laboratory, Air Force Research Laboratory Success Stories, 1997/98, "Chemical Oxygen-Iodine Laser Sets World Record for Laser Power Delivered Through a Fiber Optic," p. 14; Major John

DelBarga, et al., "Advances in Chemical Oxygen-Iodine Laser Technology, " in AFRL Technology Horizons," December 2001, pp. 22–23; "Application of High-Power Chemical Oxygen-Iodine Laser Technology," in Phillips Laboratory Success Stories, 1995, p. 61.

51. K.A. Truesdell, C.A. Helms, and G.D. Hager, "A History of COIL Development in the USA," in Proceedings, Tenth International Symposium on Gas Flow and Chemical Lasers, SPIE Volume 2502, p. 230.

52. "Airborne Laser: Program Milestones" [Online]. Available: HYPERLINK "http://www.airbornelaser.com/schedules/milestones.html" http://www.airbornelaser.com/schedules/milestones.html, July 25, 2003; "Team ABL Proposes Airborne Laser Weapon System' [Online]. Available: http://www.airbornelaser.com/news/1996/070996.html, July 28, 2003; "Boeing, Lockheed Martin, Northrop Grumman win Airborne Laser Contract" [Online]. Available: HYPERLINK "http://www.airbornelaser.com/news/1996/111296.html" http://www.airbornelaser.com/news/1996/111296.html, July 28, 2003; "Northrop Grumman Approved to Begin Manufacturing First Laser Hardware for Airborne Laser System" [Online]. Available: http://www.airbornelaser.com/news/1997/031097/html, July 28, 2003; "'First Light' Produced for Airborne Laser" [Online]. Available: http://www.airbornelaser.

com/news/1998/061598b.html, July 28, 2003; "Airborne Laser Produces 110 Percent Power" [Online] http://www.airbornelaser.com/news/1998/091698.html, July 28, 2003; "Airborne Laser Team Completes Successful Tests of High-Power Laser, Prepares to Deliver First Flight Laser Module" [Online]. Available: HYPERLINK "http://www.airbornelaser.com/news/2002/032902.html" http://www.airbornelaser.com/news/2002/032902.html, July 31, 2003.

53. McDermott, "Historical Perspective," 4; Briefing, Major Hasen, AFWL/ARBI, "Chemical Oxygen-Iodine Lasers (COIL)," May 22, 1986; Briefing, Major Hasen, AFWL/ARBI, "COIL Technology Status," February 13, 1987.

54. "Laser aircraft contract goes to Boeing," Kirtland Focus, November 15, 1996, pp. 1–4; "Airborne Laser: Program Milestones," [Online]. Available: http://www.airbornelaser.com, July 10, 2003.

55. Video, produced by Boeing, "Airborne Laser: Missile Defense For A Global Threat," n.d.

56. Ibid.

57. Ibid.; "Overview: Airborne Laser," [Online]. Available: HYPERLINK "http://www.airbornelaser.com" http://www.airbornelaser.com, July 10, 2003.

Acronyms

ABCS - Army Battle Command System

ACC - Air Combat Command

ACP - Assault Command Post

ACTD - Advanced Concept Technology Demonstration

ADA - Air Defense Artillery

ADCSP - Advanced Defense Communications Satellite Program

AEF - Air Expeditionary Force

AFATL - Air Force Armament Laboratory

AFCRL - Air Force Cambridge Research Laboratories

AFLC - Air Force Logistics Command

AFRL - Air Force Research Laboratory

AFOSR - Air Force Office of Scientific Research

AFSATCOM - Air Force Satellite Communications

AFSC - Air Force Systems Command

AFSCN - Air Force Satellite Control Network

AFWL - Air Force Weapons Laboratory

AJ - Anti-Jamming

ALSS - Advanced Location Strike System

AMSTE - Affordable Moving Surface Target Engagement

AMTI - Airborne Moving Target Indicator

AOR - Area of Responsibility

APL - Applied Physics Laboratory

ARPA - Advance Research Projects Agency

ASC - Aeronautical Systems Center

ASIC - Application Specific Integrated Circuit

ATM - Asynchronous Transfer Mode

ATO - Air Tasking Order

ATR - Automatic Target Recognition

AWACS - Airborne Warning and Control System

BC2A - Bosnia Command & Control Augmentation

BCT - Brigade Combat Team

BDA - Battle Damage Assessment

BG - Battle Group

C3I - Command, Control, Communications, and Intelligence

C4I - Command, Control, Communications, Computers and Intelligence

C4ISR - Command, Control, Communications, Computers, Intelligence, Surveillance, and Reconnaissance

C/NOFS - Communication/Navigation Outage Forecasting System

CAOC - Combined Air Operations Center

CCS - Charge Control System

CFACC - Combined Force Air Component Commander

CFD - Computational Fluid Dynamics

CIA - Central Intelligence Agency

CME - Coronal Mass Ejection

CMOS - Complimentary Metal Oxide Semiconductor

COCOMS - Combatant Commands

COMSAT - Communications Satellite

CRRES - Combined Release/Radiation Effects Satellite

CSCI - Commercial SATCOM Communications Initiative

CSEL - Combat Survivor/Evader Locator

DAMA - Demand Assigned Multiple Access

DARPA - Defense Advanced Research Projects Agency

DBS - Direct Broadcast Satellite

DCS - Decompression Sickness

DISA - Defense Information Systems Agency

DISN - Defense Information Systems Network

DISS - Digital Ionospheric Sounding System

DMSP - Defense Meteorological Satellite Program

DNSS - Defense Navigation Satellite System

DSARC - Defense Systems Acquisition Review Council

DSCS - Defense Satellite Communications System

DSP - Defense Support Program

DTS - Diplomatic Telecommunications Service

DUCA - Distributed User Coverage Antenna

EAR - Electronically Agile Radar

ECM - Electronic Counter Measures

EELV - Evolved Expendable Launch Vehicle

EHF - Enhanced High Frequency

EMSS - Enhanced Mobile Satellite Services

EO/IR - Electro-Optical/Infra-Red

ESC - Electronic Systems Center

ESD - Electronic Systems Division

FAA - Federal Aviation Administration

FLIR - Forward Looking Infrared System

FLTSATCOM - Fleet Satellite Communications

FSK - Frequency Shift Keying

GBS - Global Broadcast System

GBU - Guided Bomb Unit

GMF - Ground Mobile Forces

GMTI - Ground Moving Target Indicator

GOES - Geostationary Orbiting Environmental Satellite

GPS - Global Positioning System

HEMT - High Electron Mobility Transistor

HMMWV - High Mobility Multipurpose Wheeled Vehicle

IBS - Integrated Broadcast Services

ICBM - Inter-Continental Ballistic Missile

IDCSP - Initial Defense Communication Satellite Program

IDNX - Integrated Digital Network Exchange

IGTD - Inertial Guided Technology Demonstration

IGY - International Geophysical Year

IMS - Ionospheric Measurement System

IMU - Inertial Measurement Units

INMARSAT - International Maritime Satellite

INS - Inertial Navigation System

INTELSAT - Intelligence Satellite

IOC - Initial Operational Capability

IP - Internet Protocol

IRBM - Intermediate Range Ballistic Missile

IRTSS - Infrared Target-Scene Simulation Software

ISOON - Improved Solar Optical Observing Network

ISR - Intelligence, Surveillance, and Reconnaissance

IUS - Interim Upper Stage

JASSM - Joint Air-to-Surface Standoff Missile

JBS - Joint Broadcast Service

JDAM - Joint Direct Attack Munitions

JRSC - Jam Resistant Secure Communications

JSOW - Joint Stand-Off Weapon

JSTARS - Joint Surveillance Target and Attack Radar System

JTIDS - Joint Tactical Information Distribution System

JWID - Joint Warrior Interoperability Demonstration

LAAS - Local Area Augmentation System

LAN - Local Area Network

LANTIRN - Low Altitude Navigation and Targeting Infrared for Night

LDR - Low Data Rate

LEASAT - Leased Satellite

LEO - Low Earth Orbit

LES - Lincoln Experimental Satellites

LGB - Laser Guided Bomb

LMST - Light Multi-band Satellite Terminals

LNA - Low-Noise Amplifier

LORAN - Long Range navigation

LOS - Line of Sight

LSI - Large Scale Integrated

MBA - Multiple-Beam Antennas

MCS - Master Control System

MDR - Medium Data Rate

MEF - Marine Expeditionary Forces

MEMS - Micro Electro Mechanical Systems

MERA - Molecular Electronics for Radar Applications

MILDEP - Military Department

MILSATCOM - Military Satellite Communications

MMIC - Monolithic Microwave Integrated Circuit

MSM - Magnetospheric Specification Model

MSP - Milstar Spacecraft Processor

MTBF - Mean Time Between Failures

MTE - Moving Target Exploitation

MTI - Moving Target Indicator

NACA - National Advisory Committee for Aeronautics

NASA - National Aeronautics and Space Administration

NASP - National Aero-Space Plane

NAVSPASUIR - Naval Space Surveillance System

NCSE - Node Center Support Element

NDGPS - National Differential GPS

NEAR - Near Earth Asteroid Rendevouz

NIDL - National Information Display Laboratory

NIMS - Navy Ionosphere Monitoring System

NOAA - National Oceanic and Atmospheric Administration

NORAD - North American Air Defense Command

NRL - Naval Research Laboratory

NRO - National Reconnaissance Office

OCD - Operational Concept Demonstration

OPLANS - Operation Plans

ORT - Overland Radar Technology

OSO - Orbiting Solar Observatory

OTH - Over-The -Horizon

PDM - Presidential Decision Memorandum

PGM - Precision Guided Munitions

PoP - Points-of-Presence

PPB - Positive Pressure Breathing

PPS - Partial Pressure Suit

PRISM - Parameterized Real-Time Ionospheric Specification Model

PRN - Pseudo Random Noise

RADC - Rome Air Development Center

RASSR - Reliable Advanced Solid State Radar

RF - Radio Frequency

RLV - Reusable Launch Vehicle

RSO - Reconnaissance Systems Officer

RSOI - Reception, Staging, Onward movement, and Integration

RSTN - Radio Solar Telescope Network

SAC - Strategic Air Command

SAM - Surface-to-Air Missile

SAMSO - (Air Force) Space and Missile Systems Organization

SAR - Synthetic Aperture Radar

SAS - Space Aggressor Squadron

SATCOM - Satellite Communications

SCAMP - Single Channel Anti-Jam Man Portable

SCATHA - Spacecraft Charging at High Altitudes

SCINDA - Scintillation Network Decision Aid

SCORE - Signal Communication by Orbiting Relay Equipment

SDB - Small Diameter Bomb

SEN - Small Extension Node

SEON - Solar Electro Optical Network

SHF - Super-High Frequency

SIPRNET - Secret Internet Protocol Router Network

SLEP - Satellite Life Enhancement Program

SLV - Space Launch Vehicle

SMART-T - Secure Mobile Anti-Jam Reliable Tactical Terminal

SMEI - Solar Mass Ejection Imager

SMV - Space Maneuver Vehicle

SOF - Special Operations Forces

SOON - Solar Optical Observing Network

SRAM - Static Random Access Memory

SSPA - Solid State Phased Array

STEP - Standard Tactical Entry Point

SWA - Southwest Asia

TAC - Tactical Air Command

TACAMO - Take Charge and Move Out

TAV - Trans-Atmospheric Vehicle

TCC - Time Critical Target

TOC - Tactical Operation Center

TVRO - Television Receive Only

TWT - Traveling-Waves Tube

UAV - Unmanned Aerial Vehicle

UFO - UHF Follow-On

USAFE - United States Air Forces in Europe

USCENTCOM - United States Central Command

USCG - United States Coast Guard

USEUCOM - United States European Command

VTC - Video Teleconferencing

WAAS - Wide Area Augmentation System

WDD - Western Development Division

WSPG - White Sands Proving Grounds

Authors

Dr. J. Douglas Beason is Director of Proliferation Detection and Monitoring, Los Alamos National Laboratory, responsible for all space, intelligence, remote sensing and directed energy. A Fellow of the American Physical Society, he has served on the White House staff under the President's Science Advisor, and has experience ranging from designing nuclear weapons to commanding the USAF's Phillips Research Site. A USAF Academy graduate, he has authored 12 books, over 10 publications, is a distinguished graduate of ICAF and is a Nebula Award finalist.

Maj Gen Robert F. Behler, USAF, Ret. is the Business Area Executive for Strike Warfare, Johns Hopkins University, Applied Physics Laboratory. He has commanded the 31st Test and Evaluation Squadron, the 9th Operations Group, the 9th Reconnaissance Wing, and the Air Force Command and Control, and Intelligence, Surveillance and Reconnaissance Center. The general is a command pilot and experimental test pilot with more than 5,000 flying hours in 65 types of aircraft, including the SR–71 and U–2.

George W. Bradley III is the Command Historian at Headquarters Air Force Space Command, Peterson AFB, Colorado, a position he assumed in April 1992. He graduated from Canisius College, Buffalo, New York, with a B.A. in History in 1972 and in 1975 was awarded an M.A. in American Military History by The Ohio State University, Columbus, Ohio. The author of a number of published works on Air Force history, he began his career as a civilian Air Force historian in 1979 and has held a variety of positions.

Lt Gen Bruce K. Brown, USAF, Ret., is former Commander, Alaskan Air Command. He holds a Bachelor of Science Degree from the United States Naval Academy and Masters Degree in International Relations from George Washington University, Washington, D.C. He is a command pilot with more than 5,000 flying hours. He retired from active duty on 1 September 1985.

Dr. Russell R. Burton is a retired member of the scientific and professional cadre of senior executives as a chief and senior scientist of the School of Aerospace Medicine, Armstrong Laboratory, and Air Force Research Laboratory. His is research expertise is acceleration physiology focused on the development of anti-G protective methods/systems for fighter pilots. He has authorednumerous scientific publications and received many awards and medals for his contributionsto the scientific and Air Force communities.

Dr. Robert R. Butts received his PhD in mathematics in 1969, he began his career at the Laser Division of the Air Force Weapons Laboratory (now part of the AFRL). He has specialized in the areas of laser beam propagation through the atmosphere and adaptive optics. For the past ten years, he has been the technical lead of AFRL's ABL Technology Program for which he received the Department of Defense Distinguished Civilian Service Award in 2001.

Major William B. Cade entered the Air Force through the ROTC program at Texas A&M University (B.S. in physics) in 1987. Since then he has provided space weather support to NORAD and US Space Command missions, worked at the Air Force Space Command Headquarters, and served as an Air Force ROTC instructor. The Air Force sponsored him in obtaining a Master's Degree and then Ph.D. in physics from Utah State University. He is currently assigned to the Air Force Weather Agency Chief of the Applied Technology Division.

Brig Gen Daniel J. Darnell is Commander, Space Warfare Center, Air Force Space Command, Schriever Air Force Base, Colo. The mission of the Space Warfare Center (SWC) is to advance Air Force, joint and combined space warfare through innovation, testing, tactics development and training. General Darnell has served on the Joint Staff in Washington, D.C., and has commanded the "Thunderbirds." He is a command pilot with more than 4,500 flying hours.

John H. Darrah retired from the federal government as a member of the Senior Executive Service and Chief Scientist of Space Command in the fall of 1999. He continues to consult at a high level of government. John is nationally recognized in the many missions and systems of the Space Commands and has been a principle contributor from the cold war's missile warning and space surveillance to creating today's many new space capabilities. His work to make survivable the United State Strategic Forces and their command and control has been recognized nationally and internationally. Mr. Darrah has advised many government agencies on nuclear weapons, their effects and alternative systems designs. John served as an advisor on the Strategic Arms Limitation Talks, and was loaned to the Reagan administration during its transition period in 1980-81 to, among other things, work on developing the national security budget for the new administration. Mr. Darrah is an Adjunct Staff Member Institute for Defense Analysis (IDA) Washington D.C. (2000-to the Present) IDA is a Federally Funded Research and Development Center (FFRDC). Mr. Darrah Supports the Commander in Chief (CINC) North; Commander of Air Force Command Command, Asst. Sec. for Defense OSD/C3I, and other DOD and Government Offices and Agencies at a high level. Mr. Darrah holds a Bachelor of Arts and a Master of Science degree in Physics University of Nevada, Reno

W. Denig is a space physicist within the Battlespace Environment Division, Space Vehicles Directorate, Air Force Research Laboratory. He is the author of numerous articles dealing with the space environment and the effect of space on military systems. He holds a Masters Degree in Nuclear Science and a PhD in space physics from Utah State University.

Robert W. Duffner is a historian at the Air Force Research Laboratory's Historical Information Office at Kirtland Air Force Base, New Mexico He received his PhD from the University of Missouri and is the author of Airborne Laser: Bullets of Light (1997) and Science and Technology: The Making of the Air Force Research Laboratory (2000). He is currently working on a book on the history of adaptive optics in the Air Force.

General Ronald R. Fogleman, USAF, Retired, is currently the President and Chief Operating Officer of Durango Aerospace Incorporated. General Fogleman is the formerChief of Staff of the United States Air Force. Prior to that, he managed the Air Mobility Command and served as Commander and Chief, US Transportation command. He is currently a member of the NASA Advisory Council, NASA's Return to Flight Task Force and Defense.

Dr. Ivan A. Getting, a founding member of the Air Force Scientific Advisory Board in 1946,contributed extensively to the development of radar and military systems for command and control during the 1950s. From 1960-1977 he served as the first president of the Aerospace Corporation. His conceptual work on a space-based guidance system for rail-mobile ICBMs and subsequent advocacy of space-based navigation were critical to the development of the Global Positioning System. This was his last paper. He died on October 11, 2003.

Jon S. Jones is Senior Scientist, Air Force Research Laboratory (AFRL), Rome Research Site, Rome, New York. He holds Bachelors of Science and Masters of Science Degrees in Electrical Engineering from the State University of New York at Buffalo and Syracuse University respectively. His field of research is data fusion for Ground Moving Target Indicator (GMTI), in particular its application to Joint STARS.

Lt Gen Daniel P. Leaf, a native of Shawano, Wisconsin, is Vice Commander of Air Force Space Command. He has commanded a flight, two squadrons, an operations group, and two fighter wings, and has directed joint operations. He commanded the 31st Air Expeditionary Wing comprised of F–16, F–15, F–117, A–10, and EC–130 aircraft during Operation Allied Force. During Operation Iraqi Freedom, he served as the Director, Air Component Commander in Kuwait including Allied Force, Joint forge, Northern Watch and Southern Watch combat missions.

Dr. Alexander H. Levis is University Professor of Electrical, Computer and Systems Engineering currently on leave from George Mason University serving as the Chief Scientist of the Air Force. He joined GMU in 1990 and served twice as chair of the Systems Engineering department. He is Fellow of IEEE, AAAS, and an Associate Fellow of AIAA and has served as president of the IEEE Control Systems Society. He was educated at MIT where he also worked for eleven years as a senior research scientist.

Dr. Ruth P. Liebowitz is currently Chief of the History Office at the Electronic Systems Center, Hanscom Air Force Base, Massachusetts. Her research interests include the history of radar and reconnaissance since World War I, and also military sponsorship of science and technology.She holds a doctorate in History and a master's in History of Science from Harvard University.

Dr. Robert J. Mailloux is Senior Scientist, Sensors Directorate, Air Force Research Laboratory, Hanscom Air Force Base, Massachusetts. He is the author or co-author of numerous journal articles. Mailloux holds a B.S. Degree from Northeastern University and M.S. and Ph.D. degrees from Harvard University.

Lt Gen Forrest McCartney retired from the Air Force in 1987. During his Air Force career, he held positions as Program Director for several major satellite programs, Commander of the Ballistic Missile Organization and Commander of Space Division. In late 1986 he was assigned as the Director of Kennedy Space Center, a post he held until his retirement from the Air Force in 1987. He remained with NASA as a civilian and continued as Director of KSC until 1991. In 1994 he joined Lockheed Martin as the Astronautics Vice President for Launch Operations. He retired from Lockheed Martin in 2001.

Dr. George C Mohr, a Chief Flight Surgeon, served 39 years with the USAF in a variety of clinical, command and scientific posts. Dr. Mohr, a Rhodes Scholar, received his M.D. degree from Harvard. He is board certified in Aerospace Medicine, has authored 55 scientific papers and has received numerous awards and decorations from the USAF and civilian institutions. He currently resides with his wife, Annabel, in San Antonio, TX.

Bradford W. Parkinson is Edward C. Wells Professor of Aeronautics and Astronautics (Emeritus) at Stanford University. As an Air Force colonel, he was the first Program Manager of GPS, credited with selling the concept, synthesizing the design and building the phase one system. He flew combat missions in Laos during the Vietnam War, was an instructor at the USAF Test Pilot, School, and head of the Astronautics Department at USAFA. Later he was a Vice President at Rockwell International. As a tenured Professor at Stanford, he pioneered many civil applications of GPS. He is the winner of the prestigious Draper Prize of the National Academy of Engineering.

 **Lt Gen Harry D. Raduege, Jr.,** is director, Defense Information Systems Agency, Arlington, Virginia. As director, he leads a worldwide organization of over 8,200 military and civilian personnel. This organization plans, develops, and provides interoperable command, control, communications computers (C4), and information systems to serve the needs of the President, Secretary of Defense, Joint Chiefs of Staff, the combatant commanders, and other Department of Defense components under all conditions ranging from peace through war. He has worked his entire career in the areas of C4 space, and information operations serving in command, operations, maintenance, engineering, plans, budgeting and readiness positions. Prior to assuming his current position, he was the director of command control systems, Headquarters North American aerospace Defense Command and US Space Command, and director of communications and information, Headquarters Air Force Space Command. He also served as the Chief Information Officer for all three commands.

 **Dr. William Sears** is an internationally recognized authority in the field of Aerospace Physiology and Aircrew Life Support Systems. For the fourteen years prior to his retirement from the USAF in 1981, he was intimately involved in research and development activities at the USAF School of Aerospace Medicine and as an exchange officer at the Royal Air Force Institute of Aviation Medicine. Following retirement he has consulted with over 25 Life Support and Airframer Companies during the development of the life support systems for several new aircraft.

 **Dr. Robert Sierakowski** is chief scientist, Munitions Directorate, Air Force Research Laboratory, Eglin Air Force Base, Fla. In this capacity he provides scientific guidance and advice to the director of the Munitions Directorate as well as the commander, Air Force Research Laboratory, on research plans and programs in conventional weapon system and armament basic research.

 **Maj Gen Joseph P. Stein** is Director of Aerospace Operations, Headquarters Air Combat Command. As such he directs operations, planning, and command and control. General Stein holds a Bachelor of Science Degree from the United States Air force Academy and is a master navigator with more than 3,200 flying hours in the B–52G, T–43, B–1B, and E–8C.

 **Dr. Rick W. Sturdevant** is Deputy Command Historian at HQ, Air Force Space Command; he joined the Air Force History Program in 1984. He has authored numerous reviews, articles and papers on military space history that have appeared in various scholarly journals and books. He has also presented papers on military space history at numerous conferences and symposia. An active member of the American Astronautical Association, he is also a member of a number of other professional historical organizations.

Dr. Thomas W. Thompson heads the Office of History, Air force Research laboratory (AFRL), Rome Research Site, Rome, New York. Dr. Thompson is an Air Force veteran, having served as a radar weapons controller from 1967 until 1971. He holds a Ph.D. in history from Miami University, Oxford, Ohio, and has studied the history of science at Cornell University.

Ronald E. Thompson retired from the Aerospace Corporation in July 2002 after spending 43 years in the aerospace industry; the last 38 years at the Aerospace Corporation. Mr. Thompson joined Aerospace in 1964 as a member of the Gemini Launch Vehicle program Office and participated in all the Gemini launches as part of a joint Air Force/NASA launch team. After several years as a member of the Manned Orbiting Laboratory Program he held various management positions working to integrate military satellites for flight on NASA's Space Shuttle. In 1991, Mr. Thompson added the Military Space Test Program to his responsibilities. From 1993 to his retirement in 2002 Mr. Thompson was a member of the Milsatcom organization and was responsible for supporting the Air Force as a member of the management team for the DSCS and Milstar programs. When he retired Mr. Thompson had participated directly in 22 successful military space missions.

Dr. Van Trees is a graduate of West Point and M.I.T.. He served on the M.I.T. faculty as a Professor of Electrical Engineering. While at M.I.T., he wrote the 3-volume series on Detection and Estimation Theory, which is a classic in the field. He has been involved with satellite communications for over 30 years. As Chief Scientist at DCA, he formed the first Military Satellite Office, which generated the first comprehensive milsatcom architecture. He directed the Advanced Planning group at Comsat for 3 years and edited a comprehensive book on Satellite Communications. As ASD(C3I), he imitated the Milstar program . At M/A-COM, he built the first set of transportable SHF terminals for the Air Force. He taught a classified course on Military Satellite Communications for AFCEA for over 20 years. He is currently a Professor of Electrical Engineering and Director of the C3I Center at George Mason University. He recently finished a comprehensive book on Optimum Array Processing which is Part IV of the DEMT series.

Dr. Edward A. Watson is the Technical Advisor, Electro-Optical Technology Division, Air Force Research Laboratory and has been involved in research in active and passive electro-optic sensors. He is also an Adjunct Assistant Professor of Electrical Engineering at the Air Force Institute of Technology. He has a Ph. D. in Optics (University of Rochester), and an M. S. in Optical Sciences and a B. S. in Physics (University of Arizona). He is a Fellow of the SPIE.

Dr. Billy E. Welch. Prior to his retirement from Federal Civilian Service, Dr. Welch was thefounding Director of the Air Force's Armstrong Laboratory. Earlier assignments were in the USAF School of Aerospace Medicine, in the Human Systems Division as Chief Scientist and in the Office of the Secretary of the Air Force as the first Special Assistant for Environmental Quality. He currently is involved in S & T consulting and is affiliated with the Los Alamos National Laboratory.

Dr. Robert P. White is retired from the U. S. Air Force, where he served for ten years as a Signals Intelligence Officer in positions that included squadron operations officer and squadron commander. Following graduation from Ohio State University with an Air Force sponsored degree in military history, he served as an Air Force historian and Chief of Air Staff History. His biography on Maj Gen Mason Patrick was published by the Smithsonian Institution Press in 2001. He is currently the historian for the Air Force Office of Scientific Research in Arlington, Virginia.

Maj Gen Stephen G. Wood is Commander, Air Warfare Center, Nellis AFB, Nevada. A command pilot with over 3,400 hours, he leads the Air Force's premier training and testing center, including the Air Force's only operational remotely piloted aircraft squadrons. He served as Commander, Combined Air Operations Center, Prince Sultan AB, Saudi Arabia, providing critical command and control during Operation ENDURING FREEDOM.

Dr. Michael I. Yarymovych is President of the International Academy of Astronautics since 1996. Until 2000 he was the Chairman of the NATO Research and Technology Board, and prior to that was Chairman of AGARD. In February 1998 Dr. Yarymovych retired from Boeing, and its predecessor Rockwell International, where he held various positions as Vice President responsible for engineering and technology on Missile Defense, GPS, Space Shuttle, B1 aircraft and classified programs. He was Chief Scientist of the U. S. Air Force, Deputy for Requirements to the Assistant Secretary of the USAF for R&D and Technical Director of the USAF Manned Orbital Laboratory.

www.ingramcontent.com/pod-product-compliance
Lightning Source LLC
Chambersburg PA
CBHW080801180526
45168CB00006B/2293

9781508687337